T0254501

The Philosophy of Evolutionary Theory

Natural selection, mutation, and adaptation are well-known and central topics in Darwin's theory of evolution and in the twentieth- and twenty-first-century theories which grew out of it, but many other important topics are used in evolutionary biology that raise interesting philosophical questions. In this book, Elliott Sober analyzes a much larger range of topics, including fitness, altruism, common ancestry, chance, taxonomy, phylogenetic inference, operationalism, reductionism, conventionalism, null hypotheses and default reasoning, instrumentalism versus realism, hypothetico-deductivism, essentialism, falsifiability, the principle of parsimony, the principle of the common cause, causality, determinism versus indeterminism, sensitivity to initial conditions, and the knowability of the past. Sober's clear philosophical analyses of these key concepts, arguments, and methods of inference will be valuable for all readers who want to understand evolutionary biology in both its Darwinian and its contemporary forms.

ELLIOTT SOBER is Hans Reichenbach Professor Emeritus and William F. Vilas Research Professor Emeritus of Philosophy at the University of Wisconsin–Madison. His previous publications include *Ockham's Razors: A User's Manual* (Cambridge, 2015) and *The Design Argument* (Cambridge, 2018).

The Philosophy of Evolutionary Theory

Concepts, Inferences, and Probabilities

ELLIOTT SOBER
University of Wisconsin–Madison

Shaftesbury Road, Cambridge CB2 8EA, United Kingdom

One Liberty Plaza, 20th Floor, New York, NY 10006, USA

477 Williamstown Road, Port Melbourne, VIC 3207, Australia

314–321, 3rd Floor, Plot 3, Splendor Forum, Jasola District Centre, New Delhi – 110025, India

103 Penang Road, #05–06/07, Visioncrest Commercial, Singapore 238467

Cambridge University Press is part of Cambridge University Press & Assessment, a department of the University of Cambridge.

We share the University's mission to contribute to society through the pursuit of education, learning and research at the highest international levels of excellence.

www.cambridge.org
Information on this title: www.cambridge.org/9781009376051

DOI: 10.1017/9781009376037

© Elliott Sober 2024

This publication is in copyright. Subject to statutory exception and to the provisions of relevant collective licensing agreements, no reproduction of any part may take place without the written permission of Cambridge University Press & Assessment.

First published 2024

A catalogue record for this publication is available from the British Library

A Cataloging-in-Publication data record for this book is available from the Library of Congress

ISBN 978-1-009-37605-1 Hardback
ISBN 978-1-009-37601-3 Paperback

Cambridge University Press & Assessment has no responsibility for the persistence or accuracy of URLs for external or third-party internet websites referred to in this publication and does not guarantee that any content on such websites is, or will remain, accurate or appropriate.

for Norma

Contents

Figures

Tables

Preface

This book is about basic concepts, important arguments, and methods of inference in evolutionary biology. In each chapter I use Darwin's theory as a point of departure and then selectively take up ideas from twentieth- and twenty-first-century biology that I think are philosophically interesting. The book is not a history of evolutionary theory.

My focus on concepts involves questions like the following:

- What is fitness?
- What is a unit of selection?
- What is drift?
- What is random mutation?
- What is evolutionary gradualism?
- What is adaptation?
- What is phylogenetic inertia?
- What is a species?
- What is race?

I also assess the validity of arguments. For example:

- Do the definitions of altruism and selfishness used in evolutionary biology entail that natural selection cannot cause altruism to evolve?
- Do assumptions about random mutation and natural selection show that evolutionary gradualism is true?
- Does using a principle of parsimony in phylogenetic inference require the assumption that the evolutionary process proceeds parsimoniously?

When I consider the definitions of concepts, my goal is not to uncritically record what various scientists have said. Rather, my focus is on how these concepts ought to be understood; my project is *normative*, not purely *descriptive*. Biologists often assign different meanings to a concept, their characterizations of concepts are sometimes vague, and they sometimes fail to grasp the implications of what they propose. My goal is to sharpen concepts so that they better serve the goals of theorizing. Here I find myself drawn to Rudolf Carnap's concept of *explication*, which he used to label the ameliorative ambitions of his philosophical projects. Carnap (1950) took his explications to be proposals that should be appraised for their usefulness; for Carnap, the wrong question to ask is whether an explication is true. Unlike Carnap, I do not balk at making claims about the truth of proposed conceptual analyses, my own included.

I also do not blush when I supply Darwin's various theses with clarifications and arguments that he never considered. When I do so, I am not wearing the historian's hat. Rather, I am evaluating the *propositions* that Darwin endorsed; these propositions are separable from the *thought processes* that led him to embrace those propositions. This is not anachronistic. What would be anachronistic is attributing those clarifications and arguments to the man himself.

Interspersed with my discussion of various biological issues, I consider broader ideas from philosophy of science, such as:

- operationalism
- reductionism
- conventionalism
- null hypotheses and default reasoning
- instrumentalism versus realism
- Duhem's thesis
- hypothetico-deductivism
- the underdetermination of theory by evidence
- essentialism
- falsifiability
- the principle of parsimony
- the principle of the common cause
- interpretations of probability
- what it means for one cause of an event to be more powerful than another
- determinism versus indeterminism
- the knowability of the past

Philosophy of biology is a subject in its own right but it should not be isolated from the rest of philosophy of science.

Probability is a thread running through this book and it has two strands. First, there is the probabilistic character of evolutionary theory itself, which I discuss in several contexts – the definition of fitness, the bearing of Simpson's paradox on the evolution of altruism, the neutral theory of molecular evolution, the concept of random mutation, and Markov models of evolution. Second, there is the use of probability to make inferences about various propositions in evolutionary biology. Bayesianism is an influential epistemology in philosophy of science, but it is not the one I favor. I am more inclined to reach into a grab-bag that contains the Law of Likelihood and the Akaike information criterion. As a result, I address topics in the philosophy of probability and statistics. Instead of starting the book by describing all the probability tools I want to use, I introduce them gradually, as needed, in different chapters. If you want to begin with a brief probability primer, I suggest you look at the ABCs in Sober (2015, pp. 64–84) or Sober (2018, pp. 7–23).

I feel obliged to issue a content warning: In a few places in this book, I discuss ideas that grew out of sexist, classist, or racist assumptions. Understanding something of that history helps one understand ideas in biology that are neither sexist nor classist nor racist. Airbrushing history would remove painful reminders, but I think the reminders are valuable in spite of the pain.

I have placed bold-faced **questions** and **exercises** in footnotes. I hope they will be thought-provoking.

Madison, Wisconsin
July, 2023

Acknowledgments

I started work on the book you see before you with the idea of revising my book *Philosophy of Biology* (published in 1993 by Westview Press and now in print with Taylor & Francis), but along the way, numerous changes – both major and minor – accumulated. I came to think that the result is a new book, not a revision of an old one, though this raises thorny questions about identity through time that I hope experts on the Ship of Theseus can resolve. I am grateful to Taylor & Francis for allowing me to reprint excerpts and illustrations from that ancestral volume.

I thank the following people for helping me think through ideas: David Baum, Matthew Barker, John Beatty, Dylan Beshoner, Brianna Bixler, Michael Buckner, Eddy Keming Chen, Rachel Cherian, Phnuyal Abiral Chitrakar, Hayley Clatterbuck, Nathan Cofnas, Marisa Considine, Andrew Cuda, Katie Deaven, Michael Dietrich, Ford Doolittle, Marc Ereshefsky, Marcus Feldman, Joe Felsenstein, Branden Fitelson, Malcolm Forster, Hunter Gentry, Clark Glymour, Aaron Guerrero, Paula Gottlieb, Hank Greely, Daniel Hausman, Masami Hasegawa, Jonathan Hodge, Hanti Lin, Kate Lohnes, Josephine Lovejoy, Jun Li, Masahiro Matsuou, Matthew J. Maxwell, Kiel McElroy, James Messina, Alexander Michael, Roberta Millstein, Yusaku Ohkubo, Samir Okasha, Steven Orzack, Diane Paul, Bret Payseur, Erik Peterson, Abiral Chitrakar Phnuyal, Aaron Peace Ragsdale, Sean Rice, Alex Rosenberg, Casey Rufener, Alex Sardjev, Larry Shapiro, Nathaniel Sharp, Maxim Shpak, Alan Sidelle, Quayshawn Spencer, Mike Steel, Robert Streiffer, Naoyuki Takahata, Joel Velasco, Donald Waller, David Sloan Wilson, Polly Winsor, Rasmus Winther, James Woodward, and Shimin Zhao.

Finally, I am grateful to Hilary Gaskin, my editor at Cambridge University Press, and her assistant, Abi Sears, for their help, and to the William F. Vilas Trust of the University of Wisconsin–Madison for financial support.

1 A Darwinian Introduction

In this chapter I outline Darwin's theory of evolution, indicating which parts survive in modern biology and which have been discarded.[1]

1.1 Common Ancestry

Figure 1.1 is the only illustration that Darwin included in his 1859 book *On the Origin of Species by Means of Natural Selection, or the Preservation of Favoured Races in the Struggle for Life*. He used it to represent some of the main elements of his theory. The figure depicts a phylogenetic tree in which objects at the top of the page exist now, and as you go down the page you're going back in time, tracing items at the top to their recent and more remote ancestors. You can think of the objects at the top as current species, though Darwin's picture also applies to single organisms alive today. There are fifteen objects at the top of the figure, but they descend from only three of the objects (A, F, and I) at the bottom. If you trace those three back in time, they will eventually coalesce into a single common ancestor. Modern biology agrees with Darwin, in that the standard view now is:

> **Universal Common Ancestry:** All the organisms now on earth descended from a single common ancestor.

This thesis does not mean that life evolved from non-living materials only once. Universal common ancestry is compatible with numerous start-ups. There also is no requirement that only one of those start-ups managed to have descendants that are alive today. To see why, compare Darwin's Figure 1.1

[1] Darwin's theorizing evolved in his lifetime, on which see Ruse (1979) and Ospovat (1981). In this book, I use the term "Darwin's theory" to refer to what appeared in the first edition of the *Origin* and was further articulated in subsequent editions of that book and in his subsequent writings.

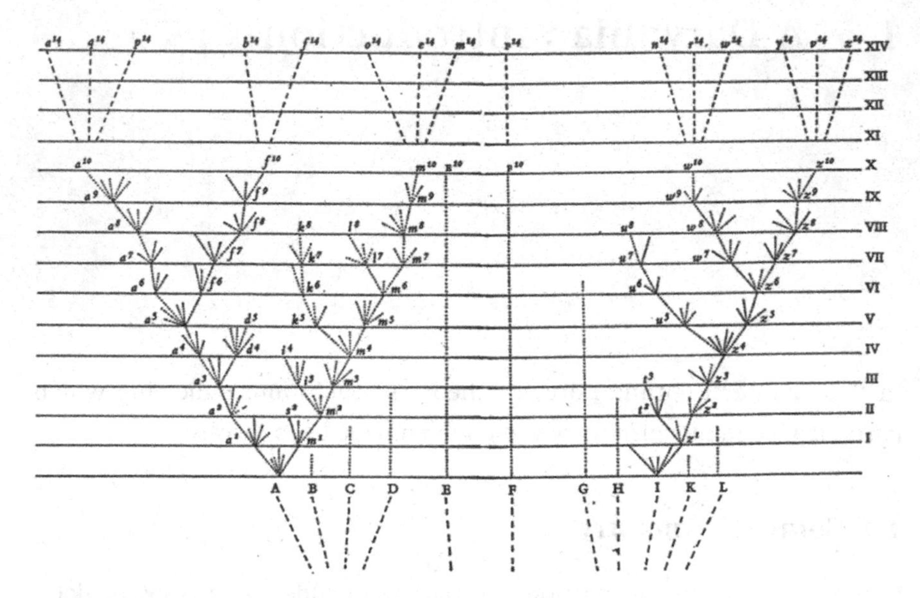

Figure 1.1 The Tree of Life in Darwin's *Origin of Species*.

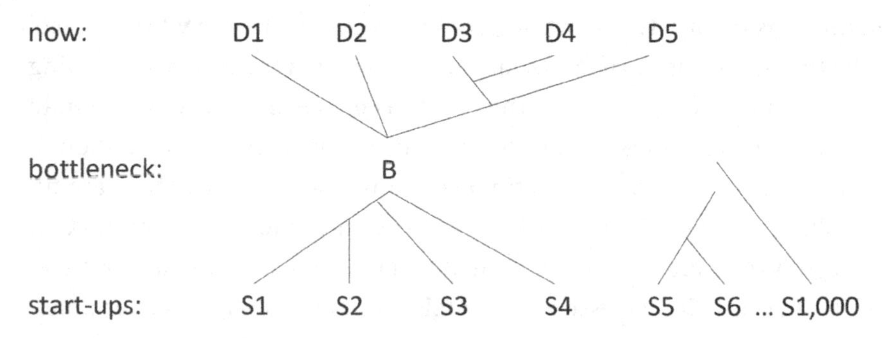

Figure 1.2 A phylogenetic bottleneck in which there is universal common ancestry.

with the genealogy shown in Figure 1.2, which includes 1,000 start-ups, four of which founded lineages that then merged; the result is a universal common ancestor of all living things on the planet today. The genealogy in Figure 1.1 is a *tree* in the technical sense of that term: Branches split but never join as your eye moves from past to present. The genealogy in Figure 1.2 is not a tree; biologists use the term "reticulation" to describe the joining.

Darwin was a resolute tree thinker, but the question of whether life's genealogy is tree-like is now debated. Some biologists regard the Tree of Life as an outmoded concept, given that pairs of plant species have often hybridized to produce new species, and given also that organisms often have genes that didn't come from their parents but were inserted by the horizontal transfer of genes from a contemporaneous organism in a different species. However, other biologists think the Tree of Life idea is true often enough, especially for eukaryotes (organisms whose cells have nuclei), for it to remain a valuable idea. See Doolittle (2000) for discussion.

1.2 Species

The title Darwin gave to his 1859 book may lead you to think that the species category occupies a very special place in his theory. If so, the following remark from the *Origin* (Darwin 1859, p. 52) may surprise you: "I look at the term species as one arbitrarily given for the sake of convenience to a set of individuals closely resembling each other, and that it does not essentially differ from the term variety." When a parent in Darwin's diagram has several descendant lineages, the descendants will count as conspecific varieties if they diverge only modestly, but if they become more different from each other, they will count as distinct species. There is no bright line that separates the one situation from the other. Darwin denies that there are "walls" that prevent old species from evolving into new ones. This no-walls thesis follows immediately from common ancestry.

Most present-day evolutionary biologists disagree with Darwin's relaxed view of the difference between species and varieties. One of the most influential species concepts in modern evolutionary biology is Ernst Mayr's (1970) *biological species concept*. It holds (roughly) that two populations belong to the same species precisely when individuals in the two populations interbreed and produce viable fertile offspring. For Mayr, subspecific varieties are different in kind from species, properly so called. I'll discuss species in Chapter 7.

1.3 Evolution

Whereas the vertical dimension in Darwin's Figure 1.1 represents time, the horizontal dimension represents phenotypic similarity.[2] Objects

[2] *Phenotype* is a garbage-can category denoting any feature of an organism that is not genetic in character; it includes all traits of morphology, physiology, and behavior.

A and B in the figure are more similar to each other than A and C are, and A and L are more dissimilar than any other pair of early objects. Biologists who now draw phylogenetic trees usually don't use the horizontal dimension in this way; in modern biology, trees represent genealogical relatedness alone.

In Figure 1.1, descendants almost always differ phenotypically from their ancestors. If evolution just means that lineages change their characteristics, then there is evolution aplenty in Darwin's diagram. But is the equation of evolution with character change correct? Suppose that nutrition improves in a starving population with the result that descendants are bigger and healthier than their ancestors. That would not be enough for modern biologists to conclude that the population has evolved. The definition often used now is that evolution is change in the genetic composition of populations.[3] Phenotypic change without genetic change isn't evolution. Unfortunately, the genetic definition is too narrow. The genetic system is itself a product of evolution. This entails that the existence of genes is not a prerequisite for evolution to occur.

What did Darwin mean when he talked about lineages evolving? Darwin (1859, p. 12) says that he restricts his attention to traits that are *inherited*. This raises the question of what inheritance means. In ordinary English, you can say that people sometimes "inherit" money from their parents, but that's not how Darwin used the term. Darwin had in mind the traits that are transmitted from parents to offspring *in the biological process of reproduction*. Gregor Mendel was a contemporary of Darwin's, but Darwin knew nothing about Mendelian genes. Modern biology reveals that the transmission of material from parents to offspring in reproduction involves more than just genes. There is non-genetic material in

This usage is compatible with the fact that many phenotypes are causally influenced by genes. The distinction between *what X is* and *what causes X* is important here. An organism's phenotype can include things that are outside its skin; the fact that a spider spins a particular type of web is a phenotypic trait the spider has. The elaboration of this idea has been the project of *niche construction theory*, which is part of what is called *the extended evolutionary synthesis* (Laland et al. 2015).

[3] This definition is broader than the definition of evolution as change in gene frequencies. The problem with the gene frequency definition is that the frequencies of single genes can remain constant while the frequencies of gene combinations change, and biologists will readily count the latter as instances of evolution. Sober (1993) illustrates this possibility by an example involving assortative mating.

both eggs and sperm. The new field of *epigenetics* studies that additional dimension of transmission.[4]

You may think that fossils provide conclusive evidence that evolution has occurred. After all, many fossils differ markedly from the organisms we now see around us. In fact, the observed features of fossils, by themselves, don't entail that evolution has occurred. However, universal common ancestry plus the fact that current organisms differ in their inherited traits suffices for evolution to have occurred. Fossils are icing on the cake.[5]

1.4 Divergence and Opportunism

Returning to the horizontal dimension of Figure 1.1, you'll notice that there is more diversity now than there was at the time of A through L. Darwin drew broken lines from A through L back into a more remote past, and he slanted those lines to indicate that there was still less diversity at that earlier time. Modern biology rejects the idea that diversity steadily increases. Mass extinctions have diminished diversity – several times, and drastically. True, when life first began there was minimal diversity, and now there is lots more. But in between, diversity has waxed and waned.

Darwin has diversity steadily increasing in Figure 1.1 because he endorsed the following principle:

> **The Principle of Divergence**: The immediate descendants of a common ancestor tend to do better at surviving and reproducing if they have extreme phenotypes, and worse if they have middling phenotypes. (Darwin 1859, p. 117)

Darwin's (1859, p. 76) idea was that individuals with extreme phenotypes usually have fewer competitors than individuals with middling phenotypes, and so the extremists do better in the struggle for existence. In other words, extremists tend to be *fitter* than middle-of-the-road organisms, and so the process of *natural selection* will favor the evolution of more extreme trait values over trait values that are middling. For Darwin, that's why diversity increases. To check whether Darwin's Figure 1.1 conforms to his Principle

[4] This expanded conception of inheritance is part of what is now called the extended modern synthesis (Laland et al. 2015).

[5] In Section 7.10, I'll explain why a fossil can provide stronger evidence than an extant organism about the traits of their most recent common ancestor.

of Divergence, you can trace the objects that exist now back in time, counting how often they derive from middling ancestors and how often from extremists.

Modern biology has abandoned this part of Darwin's theory. The Principle of Divergence is often untrue. Birth weight in humans is a classic counterexample; newborn babies of middling weight on average do better than babies who are smaller or larger. However, the problem with Darwin's principle goes deeper. It isn't just that there are exceptions to the Principle of Divergence. There is no such thing as the one and only phenotype that "tends" to evolve when there is natural selection. For example, organisms that are more complex often outcompete organisms that are simpler, but the reverse pattern is common as well. Darwin (1859, p. 148) cites an example of the latter when he notes that parasites are often simpler than their free-living ancestors. George Gaylord Simpson (1967, p. 160) expressed this idea by saying that natural selection is *opportunistic*. Which traits are fitter than which others almost always depends on accidental features of the environment.

The opportunism of evolution by natural selection contrasts starkly with the idea that evolution is *pre-programmed*. Jean-Baptiste Lamarck (1809) proposed an evolutionary theory of this type. According to Lamarck, life emerges repeatedly from non-living materials and then increases in complexity by moving through a fixed sequence of stages.[6] Lamarck's theory denies that *current* human beings and *current* worms have a common ancestor, although it asserts that *current* human beings had worm-like *ancestors*. We human beings belong to a very old lineage, since that lineage had to evolve through a long sequence of simpler stages. Current worms, being simpler than humans, trace back to a more recent start-up. Lamarck's theory denies universal common ancestry.

1.5 Selection – Natural, Artificial, Sexual, and Bi-leveled

Although the Principle of Divergence is not part of modern biology, one of Darwin's ideas about natural selection is now mainstream:

[6] This is the primary evolutionary process in Lamarck's theory, but he adds a secondary process in which organisms diversify so as to adapt to their environments (e.g., when giraffes evolve longer necks so they can reach leaves at the top of trees). This is where Lamarck's idea of the inheritance of acquired characteristics comes into play, an idea I'll discuss in Section 1.6.

Natural Selection is a pervasive and important cause of evolution.

Natural selection is Darwin's explanation for why only three of the eleven lettered objects in Figure 1.1 managed to have descendants that are alive today. Individuals that are better able to survive and reproduce outcompete individuals that are less able to do so. Natural selection causes some lineages to go extinct and others to persist. In the *Origin*, Darwin (1859, p. 6) says that "natural selection has been the main but not the exclusive cause" of evolution. Darwin's mention of "main" entails that natural selection is more important than other causes of evolution. Notice that the thesis I formulated above about natural selection is more modest. I'll explain in Chapter 8 why I've steered clear of the word "main."

Darwin distinguishes natural selection from artificial selection. Artificial selection is what farmers do when they decide which plants and animals in one generation will be permitted to reproduce. Artificial selection involves conscious choosing; natural selection, in contrast, can be mindless. If the weather gets colder, there may be selection for thicker fur in a species of bears, with the consequence that bears with thicker fur survive and reproduce more successfully than bears with thinner fur. The weather is not a conscious agent.[7]

Darwin also contrasts natural selection with sexual selection. He says that sexual selection occurs when males compete with each other for access to females, and also when females choose the males with whom they'll mate. A consequence of the former is that male elephant seals are far larger than females. A consequence of the latter is that peacocks have gaudy tails, whereas peahens do not. It is odd that Darwin thought of male combat and female choice as falling outside the category of natural selection. It also is noteworthy that Darwin didn't mention competition among females for access to males, nor did he think that males might be choosy. In his book about human evolution, *The Descent of Man*, Darwin (1871, p. 256) makes room for these possibilities when he says that sexual selection "depends on the advantage which certain individuals have over other individuals of the same sex and species solely in respect of reproduction." Darwin's idea that males are "avid" whereas females are "coy" was part of the same stereotype that led him to think that men are more intelligent whereas women are more emotionally sensitive (Hrdy 1997; Bradley 2021).

[7] **Question:** If "natural selection" just means selection that occurs in nature, should artificial selection be regarded as a kind of natural selection?

Biologists now generally view sexual selection as a kind of natural selection. In sexual selection, males compete with males and females compete with females. In each case, one sex is part of the environment that influences the evolution of traits in the other. It is facts about peahens that caused peacocks to evolve gaudy tails, and it is facts about female elephant seals that cause males to be so much larger than females.[8] Much of natural selection isn't like this – consider the example of a bear species evolving thicker fur as a response to climate change, not as a response to what members of the opposite sex are doing. Although sexual selection frequently results in a polymorphism (meaning that different individuals in the same species have different traits), non-sexual natural selection can do the same thing; I'll describe an example in Chapter 3.[9]

Another feature of Darwin's conception of natural selection is that it occurs on two different levels of organization. The first level involves individual selection, wherein organisms in the same species compete with each other. Group selection occurs when groups of conspecific organisms compete. Darwin introduced the idea of group selection because he saw that there are traits in nature that would be inexplicable if individual selection were the whole story. For example, consider what he says about the honeybee's barbed stinger (Figure 1.3):

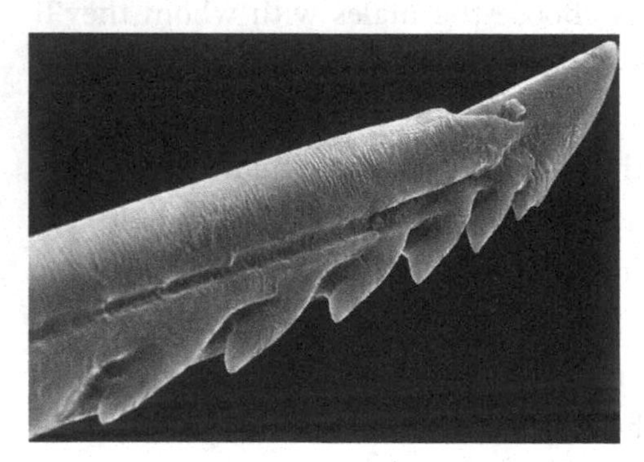

Figure 1.3 The honeybee's barbed stinger.
Source: I thank Rose-Lynn Fisher for permission to reprint "Sting 650x," a scanning electron micrograph from her book *Bee* (Princeton Architectural Press, 2012).

[8] I'll describe an example of sexual selection in dung flies impacting both sexes in Section 8.5. Here females are more "avid" than males.

[9] Sexual selection, as I have defined it, does not explain why organisms reproduce sexually; rather, it presupposes that they do.

> Can we consider the sting of the wasp or the bee as perfect, which, when used against many attacking animals, cannot be withdrawn, owing to the backward serratures, and so inevitably causes the death of the insect by tearing out its viscera?
>
> If we look at the sting of the bee, as having originally existed in a remote progenitor as a boring and serrated instrument, like that in so many members of the same great order, and which has been modified but not perfected for its present purpose, with the poison originally adapted to cause galls subsequently intensified, we can perhaps understand how it is that the use of the sting should so often cause the insect's own death: for if on the whole the power of stinging be useful to the community, it will fulfill all the requirements of natural selection, though it may cause the death of some few members (Darwin 1859, p. 202).

Bees with barbed stingers die when they sting an intruder to the nest, whereas bees with barbless stingers can withdraw their stingers without harming themselves. The barb is bad for the organism, but it is good for the group, since the barb keeps the stinger in place while it continues to pump venom after the bee has died. Groups of bees that contain individuals with barbed stingers do better at avoiding extinction and at founding new groups than groups that contain individuals with barbless stingers.

Darwin's idea of there being two levels of selection (or "units of selection") was standard during the first half of the twentieth century, but it came in for strenuous attack in the 1960s. To this day, some evolutionary biologists think that hypotheses of group selection are worse than false – they are confused and should be excluded from the toolkit of serious science – whereas other biologists now think that group selection is a coherent hypothesis, the plausibility of which needs to be evaluated on a case-by-case basis. This controversy is the subject of Chapter 3.

1.6 Heredity

For natural selection to cause the frequencies of traits in a population to change, there needs to be heredity. As mentioned, Darwin didn't know about Mendelian genes. After publishing the *Origin*, Darwin advanced his "provisional theory of pangenesis" in his 1868 book *The Variation of Animals and Plants under Domestication*. Luckily for Darwin he did not put that theory in the *Origin*; it is full of holes. What he included instead was a very modest idea, which he called

Figure 1.4 Broken lines represent the heights of parents and the heights of offspring associated with different parental pairs. The solid line represents the average offspring heights associated with different midparent heights.

> **The Strong Principle of Inheritance**: Offspring tend to resemble their parents. (Darwin 1859, pp. 5, 127, 438)

This principle does not describe a mechanism whereby parents influence the traits of their offspring. It just says (for example) that taller than average parents tend to have taller than average offspring, as depicted in Figure 1.4. The word "tends" indicates a probabilistic relationship between parents and offspring.[10]

Darwin's silence in the *Origin* about the mechanism of inheritance did not prevent his critics from noting a big problem that his theory confronted – the problem of blending inheritance. For example, suppose that each of the organisms in a population can extract 30 percent of the nutrients that are in the food they eat, and then by chance an advantageous characteristic appears in the population that permits its bearer to extract 90 percent. Darwin would want to show that natural selection can increase the frequency of this new trait so that, some generations hence, everyone extracts 90 percent. The problem is that this won't happen if there is blending inheritance. For example, suppose that when the novel 90-percenter mates with a 30-percenter, their offspring will be able to extract 60 percent. When one of those 60-percenters mates with a 30-percenter, their offspring will be a 45-percenter. After numerous further generations, the novel trait

[10] I'll introduce the concept of mathematical expectation in Section 2.2; it can be used to characterize the probabilistic tendency depicted in Figure 1.4 more precisely.

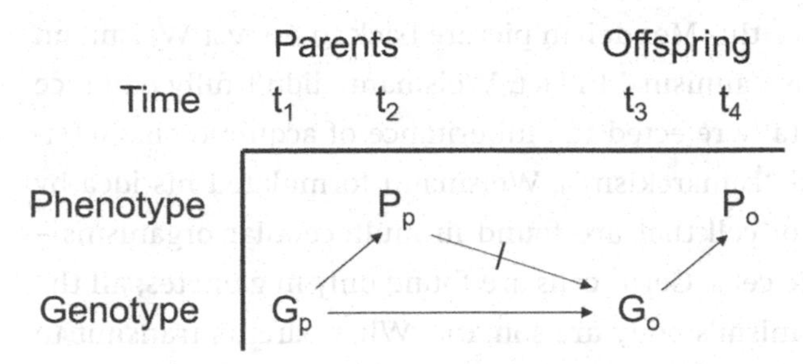

Figure 1.5 The classic Mendelian representation of development and reproduction.

will have been "diluted" to such an extent that the population will not have evolved at all. Fleeming Jenkin (1867) made this objection and it hit home. The problem was solved only after Mendel's ideas were rediscovered in the twentieth century and then integrated with Darwinian natural selection in what came to be called the *Modern Synthesis*. Mendelian genes need not blend in their phenotypic effects; it is perfectly possible that the mating of a 90-percenter with a 30-percenter should result in offspring that are all 90-percenters; this is what dominant genes do.

Mendelian genetics solved one problem for Darwin's theory, but it created another. Mendelism led to the rejection of a second principle about inheritance that Darwin endorsed. He called it

> **The Principle of Use and Disuse**: A trait acquired by a parent may be inherited from that parent by its offspring. (Darwin 1859, pp. 11, 43, 134–155, 143, 168, 447, 455, 472–473, 479–480)

Here Darwin and Lamarck were on the same page. For example, they believed that when blacksmiths working at the forge acquire big muscles, the result will be that their children have big muscles even when they do not exercise. Now the standard evolutionary view is to reject this idea by invoking the Mendelian distinction between genotype and phenotype; see Figure 1.5. The unslashed arrows represent causality, and the arrow with a slash through it denies that there is a causal connection between parental phenotype and offspring genotype.[11]

[11] Figure 1.5 fails to represent the effect of the environment on phenotype, but biologists recognize that phenotypes are almost always influenced by both genes and environment. Ignoring or denying the role of the environment is often called "genetic determinism"; it is often ridiculed, but it is sometimes a useful idealization.

Biologists often trace this Mendelian picture back to August Weismann (1892) and call it "Weismannism." In fact, Weismann didn't fully embrace Mendelism, but he totally rejected the inheritance of acquired characteristics (which he called "Lamarckism"). Weismann formulated his idea by describing two types of cell that are found in multi-cellular organisms – germ cells and somatic cells. Germ cells are found only in gametes; all the other cells in the organism's body are somatic. What parents transmit to offspring in reproduction is their germ cells, not their somatic cells, and changes in an organism's soma don't change what's in that organism's germ cells.[12]

1.7 Gradualism

Another feature of Darwin's theory that survived well into the twentieth century and still is endorsed by many biologists is his

> **Gradualism**: Evolution almost always occurs by the accumulation of a large number of small changes, not by the accumulation of a small number of large changes.

Darwin was influenced by the gradualist geology that he learned from his Cambridge teacher Charles Lyell, whose 1809 book accompanied Darwin on the Beagle voyage (Hodge 1987; Mayr 2007). However, Darwin felt the pull of a much more general idea. At the end of the *Origin*, Darwin (1859, p. 471) expresses confidence in the general principle that "*Natura non facit saltum* [nature does not make jumps], which every fresh addition to our knowledge tends to make more strictly correct." A third source for his gradualism was his detailed understanding of how plant and animal breeders do their work. In *Variation of Animals and Plants under Domestication*, Darwin (1868, p. 414) describes several varieties "that suddenly appeared in nearly the same state as we now see them," but adds that the overwhelming majority "were formed by a slow process of improvement." Darwin (1868, pp. 234–235) explains this pattern as follows: "As conspicuous deviations of structure occur rarely, the improvement of each breed is generally the result … of the selection of slight individual differences." In another passage, Darwin is

[12] **Question:** Does epigenetic inheritance contradict the Weismann doctrine depicted in Figure 1.5?

less cautious: "Slight individual differences … suffice for [both artificial and natural selection], and are probably the sole differences which are effective in the production of new species" (p. 192). Darwin's gradualism was "the received view" for most of the twentieth century, but population geneticists have found flaws in an influential argument that defends that thesis. This change isn't widely known in the rest of evolutionary biology, let alone among philosophers of biology. It's time for an update, which I'll describe in Chapter 6.

1.8 Mutation

Darwin's gradualism concerns the mutations that are apt to increase in frequency in populations once they occur. This raises the question of why mutations occur in the first place. Darwin's answer remains a pillar in modern evolutionary theory:

> **Mutations are random:** Mutations have their causes, but they do not occur because they would be useful to the organisms in which they occur.

In *Variation of Animals and Plants under Domestication*, Darwin (1868, pp. 248–249) explains this thesis by presenting a beautiful analogy:

> Let an architect be compelled to build an edifice with uncut stones, fallen from a precipice. The shape of each fragment may be called accidental; yet the shape of each has been determined by the force of gravity, the nature of the rock, and the slope of the precipice, – events and circumstances all of which depend on natural laws; but there is no relation between these laws and the purpose for which each fragment is used by the builder. In the same manner the variations of each creature are determined by fixed and immutable laws; but these bear no relation to the living structure which is slowly built up through the power of selection whether this be natural or artificial selection.

I'll discuss the meaning and justification of the thesis that mutations are random in Chapter 6.

Darwin thought that natural selection and mutation have something in common. He says in his *Autobiography* (1958, p. 50) that "there seems to be no more design in the variability of organic beings and in the action of natural selection, than in the course which the wind blows." This similarity sits side by side with a difference: Mutations do not arise because they are useful, but variants that exist in a population increase in frequency because they are useful. In both cases, the processes can and do occur without guidance from an intelligent designer.

1.9 Was Darwin "Illogical"?

In his essay "Darwin's Five Theories," Mayr (2007, pp. 97–98) says that Darwin was "illogical" because he failed to recognize that his theory was in fact five distinct theories that are logically independent of each other.[13] For Mayr, this was no small failing; he says that "Darwin's blindness to recognize this [logical independence] became one of the main reasons for the never-ending controversies on evolutionary biology after 1859." Here's how Mayr separates the parts of the Darwinian picture:

(1) evolution as such
(2) common descent
(3) gradualism
(4) multiplication of species
(5) natural selection

Although Mayr is sometimes unclear about which propositions he wants these phrases to denote,[14] his list does correspond significantly to the parts of Darwin's theory that I've described in this chapter.

I have two beefs with Mayr's charge of illogicality. First, a slightly enriched representation of Darwin's theory shows that there are logical dependencies among its elements. I noted earlier that common ancestry, conjoined with the observation that current species differ in their inherited characteristics, entails that evolution must have occurred, and that

[13] Proposition A is logically independent of proposition B precisely when the four conjunctions A&B, A¬B, notA&B, and notA¬B are all logically possible, meaning that none of them is a contradiction. Probabilistic independence is different from logical independence, as I'll explain in Chapter 6.

[14] For example, here is how Mayr (2007, p. 100) characterizes common descent: "every group of organisms descended from an ancestral species." What does Mayr mean by "group"? If the totality of organisms alive today doesn't count as a "group," Mayr's formulation fails to capture the idea of *universal* common ancestry. **Question:** "Everyone has a birthday" does not entail that there is a single day on which everyone was born. To think otherwise is to commit what I call "the birthday fallacy" (Sober 1990, pp. 42–43). How does this point apply to Mayr's characterization of common descent?

universal common ancestry entails that there were no "walls" that prevented one species from evolving into another. In Chapter 7, I'll describe how Darwin and his successors have used the fact of common ancestor to reconstruct the character states of ancestors; without common ancestry, the character states of ancestors would have been purely conjectural, as witnessed by Lamarck's theory that evolution pushes lineage to evolve from simpler to complex. In Chapter 8, I'll describe how Darwin used the idea of common ancestry to refute some seemingly plausible hypotheses about natural selection.

Second, I think that Darwin clearly saw that there are parts of his theory that are logically independent of each other. For example, he recognized the separateness of evolution and universal common ancestry. Darwin was aware that Lamarck's theory embraced the former while denying the latter. Darwin disliked Lamarck's theory, but he never accused it of being logically contradictory. Another example of Darwin's logical lucidity can be found in his discussion of the evidence that supports the different parts of his theory. Darwin (1859, p. 427) argued that neutral and deleterious similarities provide telling evidence for common ancestry, whereas adaptive similarities do not. The fact that dolphins and sharks are both shaped like torpedoes isn't strong evidence that they have a common ancestor, since one would expect large aquatic predators to evolve that shape even if they lacked a common ancestor. On the other hand, the fact that humans and monkeys both have tailbones is strong evidence for their common ancestry, since tailbones are useless to humans. Darwin's thoughts here, which I'll analyze in Chapter 4, show that he realized that common ancestry and natural selection are logically independent.

Mayr was right to separate the logically independent propositions in Darwin's theory, but their logical independence does not mean that they are mutually irrelevant. One of my goals in what follows is to show how the parts of the theory fit together. In tracing out these connections, I'll make use of ideas from modern evolutionary theory of which Darwin never dreamt.

2 Fitness and Natural Selection

2.1 Malthusian Struggle and Competition

Darwin (1958, p. 120) reports in his *Autobiography* that a decisive moment near the start of his thinking about evolution was reading Thomas Malthus's (1798) *Essay on the Principles of Population*. Darwin says he read Malthus "for amusement," but the impact was substantial, leading him to present a Malthusian argument in the first chapter of the *Origin* for the pervasiveness of natural selection (Darwin 1859, pp. 63–64).

Malthus does not discuss evolution in his essay, but he does postulate a dilemma that all living things confront, human beings included:

> Through the animal and vegetable kingdoms, nature has scattered the seeds of life abroad the most profuse and liberal hand.... The germs of existence contained in this spot of earth, with ample food, and ample room to expand in, would fill millions of worlds in the course of a few thousand years. Necessity, that imperious all pervading law of nature, restrains them within the prescribed bounds. The race of plants, and the race of animals shrink under this great restrictive law. And the race of man cannot, by any efforts of reason, escape from it. Among plants and animals its effects are waste of seed, sickness, and premature death. Among mankind, misery and vice. (Malthus 1798, Chapter 27).

Malthus held that human beings, like other organisms, inevitably have more offspring than the supply of food can sustain,[1] with the result that each species, ours included, is and always will be trapped in a dismal alternation of population growth and starvation. Darwin agreed (Vorzimmer 1969) and

[1] Malthus bases this thesis on two mathematical assumptions – that reproduction leads a population to grow multiplicatively whereas the population's food supply increases only additively – with the result that population inevitably outstrips food. Darwin (1859, p. 63) concurs.

added an observation to this sweeping picture – that the organisms in a species differ from each other in characteristics that are relevant to how successful they'll be at surviving and reproducing. Food scarcity induces a selection process in which some organisms survive and reproduce more successfully than others. If traits affecting survival and reproduction are inherited, the population will evolve, with more successful traits increasing in frequency and less successful traits declining.

Darwin (1859, p. 63) says that the Malthusian process *suffices* for natural selection: "A struggle for existence inevitably follows from the high rate at which all organic beings tend to increase." He does not say that the Malthusian scenario is *necessary*. Darwin was right to avoid that commitment, since scarcity of food is just one feature of a population's environment that can induce a selection process. For example, if organisms differ in their ability to resist disease or to evade predators, that can set a selection process in motion even when food is abundant.

Darwin, like Malthus, was struck by nature's *superfecundity*. For example, consider the fact that apple trees produce huge numbers of apples, with only a small fraction of the seeds in those apples growing into new trees. Before and after Darwin's *Origin* appeared, defenders of the design argument for the existence of God explained this fact by citing God's benevolence (Sober 2018). God wanted us to have apples, so he gave us trees that make that possible, in abundance. Darwin took God out of the picture and replaced the benevolence of God with a competition among apple trees. A tree that produces more apples outcompetes a tree that produces fewer. The observed superfecundity is due to mindless competition, not to intelligent benevolence. Apple trees are in an arms race with each other in which the giant reproductive output of apple trees is a *result* of evolution by natural selection, not a *precondition* for that process to occur. This is another reason why the Malthusian story provides only a partial picture of what evolution by natural selection involves.[2]

[2] Just as Malthus discounts the possibility that population size might shrink because people decide to have fewer children, Darwin's Malthusian argument ignores the possibility that selection might lead populations to evolve reduced reproductive rates. Malthus's pessimism is refuted by the so-called *demographic transition* (from higher birthrates and higher infant mortality to lower birthrates and lower infant mortality) that has occurred in many countries and is now occurring in others. Darwin's omission is interesting in view of what he says about group selection, a topic I'll discuss in Chapter 3.

When people think of competition in nature, they often think of predator and prey. It seems natural to say that lions compete with zebras. When a lion pursues a zebra, the lion's meal requires the zebra's death, and the zebra's escape ensures the lion's continued hunger. If one gains, the other loses. Charles Lyell (1830), whose influence on Darwin's gradualism I mentioned in the previous chapter, thought about competition in this way – it's an *inter*specific process. Darwin saw things differently. Zebras compete with other zebras in an environment that contains lions, and lions compete with other lions in an environment that contains zebras. For Darwin, competition is first and foremost *intra*specific (Herbert 1971).

Darwin's departure from Lyell was theory-driven. For Darwin, evolution by natural selection is a slow and gradual process spanning numerous generations in which some traits increase in frequency and others decline. This led him to focus on conspecific organisms. Zebras reproduce with each other and lions do the same. In consequence, the competition between lions and zebras decomposes into a pair of selection processes. In one, zebras evolve new traits; in the other, lions do the same. Zebras and lions are both organism and environment.

Darwin (1859, p. 62) says that the "struggle for existence" that occurs when there is natural selection needs to be understood in a "large and metaphorical sense." He says that "two canine animals in a time of dearth may be truly said to struggle with each other which should get food and live. But a plant on the edge of the desert is said to struggle for life against the drought." Darwin might better have talked about *two* plants that live miles apart in his second example. The point is that when two dogs fight over a piece of meat, nutrition gained by one is nutrition lost to the other. In contrast, two plants may differ in their ability to withstand a drought without its being true that water gained by one is water lost to the other. The dogs have a causal impact on each other, but the fates of the two plants are separate effects of a common cause. In the language of game theory (a twentieth-century invention that I'll discuss in the next chapter), the struggle for existence need not be a zero-sum game. Think of the difference in scoring between tennis and golf (Sober 1984). For both the dogs and the plants, and in both tennis and golf, there are winners and losers. I'll use the term "competition" to describe all four.

2.2 Fitness

Darwin took pains in the *Origin* to explain that natural selection does not require an intelligent agent who does the selecting, but he came to think that his readers often missed this important point. Alfred Russel Wallace[3] suggested a way to block that misunderstanding – that Darwin use a phrase that the philosopher Herbert Spencer (1864) coined to characterize the Darwinian theory – "the survival of the fittest." Darwin introduced this phrase in his 1868 book *Variation of Animals and Plants Under Domestication* and inserted it a year later in the fifth edition of the *Origin*.

Although "the survival of the fittest" conveys no suggestion of conscious agency, it has two drawbacks. First, it encourages the misleading impression that evolution by natural selection is all about some organisms out-surviving others. This neglects the importance of reproduction. An organism has a probability of surviving to reproductive age, and, if it reaches reproductive age, it may achieve some degree of reproductive success. The organism's *overall* fitness depends on both its *viability* and its *fertility*. These concepts need to be separated, since a trait can diminish one while augmenting the other. I described an example of this in the previous chapter: a peacock's gaudy tail reduces his probability of surviving to reproductive age, but it augments his chances of reproductive success.

The second drawback is that "the survival of the fittest" sounds to some like a tautology, meaning an empty definitional truth. This criticism is sometimes expressed by posing and answering two questions:

- Who are the fittest? Those who survive.
- Who survives? Those who are fittest.

The coupling of these question-and-answer pairs is said to be "circular," and the suggestion is then advanced (often by creationists – see Bethell 1976, for example) that the theory of evolution by natural selection isn't a genuine scientific theory at all – it's just a useless truism.

A logical point is needed here: Tautologies are statements of a certain kind. "If it is raining now, then it is raining now" is an example.

[3] Wallace independently developed a theory of evolution that overlaps considerably with Darwin's. Darwin arranged for Wallace's paper and Darwin's summary of his own theory to be delivered together to the Linnean Society in 1858 and to be published in the society's proceedings.

The statement is made true by the definitions of the concepts that it uses. But what statement about fitness are we talking about when we consider whether the survival of the fittest is a tautology? I take the relevant statement to be this:

> The X individuals are the fittest individuals in a population at time t_1 if and only if the X individuals at t_1 are the ones that survive until t_2 (where $t_1 < t_2$).

This sentence is alleged to be true by definition. If it is true, what's the problem? The problem is that fitness at one time can't explain survival later on if "fitness" just means survival (Brandon 1978, Mills and Beatty 1979). The conclusion is then drawn that fitness and therefore natural selection have no explanatory power.

One response to this challenge has been to say that fitness can be provided with an "independent criterion" (see, e.g., Gould 1977a). If you ask why some zebras are fitter than others, a substantive answer can be found by examining how they differ in their morphology, physiology, and behavior. These are the *sources* of their fitness differences (Sober 1984), and the statement that one zebra is fitter than another because the first runs faster than the second is not a tautology. However, this information about the two zebras provides no *general* criterion for what it takes for one organism to be fitter than another, and a general criterion is what you need if you are going to define fitness. The reason that one daffodil is fitter than another probably won't overlap much with the reason that one zebra is fitter than another. Philosophers describe this point by saying that fitness is *multiply realizable*. Fitness has different *physical bases* in different organisms.[4]

A better reply to the tautology problem is to point out that fitness is an *ability* or a *propensity* (Brandon 1978; Mills and Beatty 1979).[5] When one organism survives to adulthood while another does not, it may be true that the first had a higher probability of surviving, or it may be that they had identical probabilities of doing so. If the former, there is a difference in their viability; if the latter, their viabilities were identical, and the one out-survived the

[4] The concept of multiple realizability has been of great importance in philosophy of mind. In the mind–body problem, it is important to take account of the fact that mental properties (like believing that it is raining or being in pain) are multiply realizable. See Polger and Shapiro (2016) for discussion.

[5] I'll use these terms interchangeably.

other "by chance." The same point holds if one organism has more offspring than another; this may be because they differed in their fertilities, but it's entirely possible that their abilities were the same, and they differed in their reproductive output "by chance." I'll discuss what "chance" means more fully in Chapter 5 when I turn to the concept of *drift*.

A coin-tossing analogy is helpful here. Suppose you toss coin A and coin B repeatedly, recording how often each lands heads. If coin A landed heads more often than coin B, this *may* be because A's probability of landing heads is larger than B's, but it's also possible that the two coins had the same probability of landing heads. If you toss each coin twice, the fact that the first landed heads both times while the second landed heads just once is paltry evidence that the first had a higher probability of landing heads. However, if you tossed each coin 100 times, and observed that the first landed heads 76 times, while the second landed heads 38 times, that would be substantial evidence for the first coin's having the higher probability of landing heads. A coin has an ability to land heads, just as an organism has an ability to survive to adulthood. In both cases, the *ability* is distinct from the observable outcome.[6]

Equating an organism's fitness with its actual degree of success in surviving and reproducing isn't just a mistake; it's a mistake with a history. *Operationalism* is a philosophy that has influenced several areas of science. As formulated by Percy Bridgman (1927), it asserts that a theoretical concept must be defined by describing the observational procedures that dictate when the concept applies and when it does not. The basic idea is conveyed by each of the following question-and-answer pairs:

- What is an individual's intelligence? By definition, it's the individual's score on an intelligence test.
- What is an individual's temperature? By definition, it's what the thermometer says.

[6] The difference between an ability and its observable manifestation is a familiar theme in philosophical discussion of *dispositions* (Choi and Fara 2021). A lump of sugar is water-soluble, even if it is never immersed in water and so never dissolves. The disposition underwrites a counterfactual conditional: "If this lump of sugar were immersed in water (in the right way), it would dissolve." In contrast, the bias of a coin and the fitness of an organism need to be understood probabilistically. Philosophers sometimes reserve the term "propensity" for probabilistic dispositions. I'll discuss the propensity interpretation of probability in Chapter 9.

Both of these answers are mistaken. Intelligence tests and thermometers can be *imperfectly* reliable. This must be true when two procedures for assessing temperature, or for assessing intelligence, disagree. The simple distinction that needs to be drawn here is the difference between "x is evidence for y" and "x defines y." Granted, it should be possible for observations to provide evidence concerning whether a scientific concept applies, but operationalism goes beyond that unobjectionable point. See Chang (2021) for further discussion.

Although Darwin's definition of natural selection as the survival of the fittest is unfortunate, there is an intimate connection between selection and fitness that needs to be recognized. A population of organisms experiences natural selection at a given time if and only if the organisms in the population differ in fitness at that time.[7] Darwin thought about selection in this way and the idea has been a mainstay ever since.

As mentioned, the overall fitness of an organism has two components – viability and fertility. Viability is defined probabilistically – it's the organism's probability of surviving to reproductive age. Can fertility be characterized probabilistically? The standard place to begin is the idea of *mathematical expectation.* If you toss a fair coin three times, there are eight possible outcome sequences (HHH, HHT, HTH, THH, TTH, THT, HTT, TTT), and each has a probability of $\frac{1}{8}$. The expected number of heads in this three-toss experiment is 1.5, meaning (roughly) that that's the average number of heads you'd get if you did this three-toss experiment repeatedly. Why is 1.5 the expected number? Here's how to do the math: When you toss the coin three times, there's a probability of $\frac{1}{8}$ that you'll get 3 heads, a probability of $\frac{3}{8}$ that you'll get two, a probability of $\frac{3}{8}$ that you'll get 1, and a probability of $\frac{1}{8}$ that you'll get 0. The weighted average of these four possible outcomes is the expected value: $0\left(\frac{1}{8}\right)+1\left(\frac{3}{8}\right)+2\left(\frac{3}{8}\right)+3\left(\frac{1}{8}\right)=\frac{12}{8}$. Notice that the expected number of heads when you toss the coin three times is not the number of heads you should expect when you toss the coin three times.

[7] Notice that this definition is compatible with a given organism's fitness changing, which is a good thing, since the organism's environment can change. Ramsey's (2006) concept of *block fitness* rules this out. See Millstein (2009) for discussion of what a population is.

An organism's expected number of offspring is the average number of offspring it would have if it lived its life again and again in the same circumstances. If p_0 is the organism's probability of having zero offspring, p_1 is the organism's probability of having exactly one offspring and so on, then

$$\text{the organism's expected number of offspring} = (0)p_0 + (1)p_1 + (2)p_2 + \ldots + (n)p_n = \sum_i i(p_i).$$

Since an organism's overall fitness is the mathematical product of its viability (its probability of reaching reproductive age) and its fertility (its expected number of offspring if it reaches reproductive age), the organism's overall fitness is itself a mathematical expectation.[8]

Notice that I have described fertility without requiring that offspring resemble their parents.[9] This separation is mandated by a central idea in the field of quantitative genetics called the *breeder's equation*:

response to selection = intensity of selection × heritability

The intensity of selection represents how much variation in fitness there is in the population (Plomin et al. 2017). The response to selection describes how much change in trait frequency (or in the average value of a quantitative phenotype) there will be in the next generation. The response to selection is zero if the heritability is zero. Selection does not entail evolution.[10]

2.3 Fitness Refined

The definition of overall fitness in terms of viability and fertility requires two modifications. The first has to do with its one-generation time frame. Why talk only about offspring? What about grand offspring, and more remote descendants?

Sometimes you can think about the long-term effects of selection by using a one-generation time frame for the concept of fitness. If the fast zebras in each generation of a herd evade lion attacks more successfully

[8] **Question:** How does the concept of mathematical expectation apply to the relationship of mid-parent height to offspring height depicted in Figure 1.4?

[9] The two are sometimes blended together; see, for example, Brandon's (1978) principle D and Byerly and Michod's (1991) definition of fitness, the latter of which Maynard Smith (1991) criticizes.

[10] Examples of selection without evolution will be described in Sections 3.8 and 8.3.

than do their slow contemporaries, the herd will gradually evolve a faster average running speed over multiple generations. What happens in the long term is just the sum of what happens in a sequence of short terms. However, sometimes a one-generational time frame is insufficient. An interesting example in which the time frame needs to be widened is the evolution of sex ratio. Why is the human sex ratio around 51 percent males at birth, while other species have sex ratios that are strongly female-biased? To theorize about how natural selection influences population sex ratio, the usual strategy (initiated by Darwin 1871, and refined by Düsing 1884, Fisher 1930, Hamilton 1967, and others; see Sober 2011b for discussion) is to think about a parent's *sex-ratio strategy*, meaning the mix of sons and daughters that the parent produces.

How can a parent that has 5 sons and 5 daughters do better in terms of reproductive success than a parent in the same population who has 10 sons and 0 daughters, or 0 sons and 10 daughters? If fitness just means expected number of offspring, the three strategies are equally fit. Darwin's insight, made more precise by Düsing, was to think about *grand* offspring. Consider three generations in a population – the parents (G_1), the offspring (G_2), and the grand offspring (G_3). If there are m males and f females in G_2 and t individuals in G_3, then the average male in G_2 has m/t offspring in G_3 and the average female has f/t. This means that if there are more males than females in G_2, a parent in G_1 will maximize her number of grand offspring in G_3 if she produces all daughters; symmetrically, if there are more females than males in G_2, a parent in G_1 does best by producing all sons.[11]

The second modification that needs to be made in the definition of fertility fitness (and hence of overall fitness) concerns the idea that fitness is a mathematical expectation. I'll now describe two examples in which two traits begin with equal frequencies in the population, have the same expected number of offspring, but differ from each other in their *variance in offspring number*.

The first example involves *between-generation variance* in offspring number. Individuals with trait A always have 2 offspring, whereas the individuals

[11] A simpler example that also illustrates the need for thinking about grand offspring in the definition of fitness comes from our own species. If parents provide resources to their *grand*children, selection can influence the evolution of this behavior.

with trait B in a given generation all have 1 offspring or all have 3, with equal probability.[12] The expected number of offspring is 2 for both traits. Assume that individuals in one generation die after reproducing asexually, and that offspring always resemble their parents. To consider the simplest case in which the two traits differing in their variance affects their evolution, suppose there is 1 A individual and 1 B individual in the first generation. This means that there will be 2 A individuals in the next generation and either 1 B individual or 3, with equal probability. Although the two traits begin with the same population frequency and have the same expected number of offspring, their expected frequencies[13] in the next generation differ:

$$\text{Expected frequency of A in generation two} = \left(\frac{1}{2}\right)\left(\frac{2}{3}+\frac{2}{5}\right) = 0.533$$

$$\text{Expected frequency of B in generation two} = \left(\frac{1}{2}\right)\left(\frac{1}{3}+\frac{3}{5}\right) = 0.467$$

Where do these numbers come from? For trait A, there's a 50 percent chance that the population will be $\frac{2}{3}$ A and a 50 percent chance that it will be $\frac{2}{5}$ A; for B, there's a 50 percent chance that the population will be $\frac{1}{3}$ B and a 50 percent chance that it will be $\frac{3}{5}$ B. In this example, the trait with the lower variance is the one that increases in frequency (in expectation).

A second way in which variance can be relevant to fitness involves *within-generation* variance in offspring number. Gillespie (1974) provides the nice example of a bird species in which a nest of eggs has a probability of being destroyed by predators of 0.9. Suppose birds with trait C lay 10 eggs in just one nest, while birds with trait D create two nests containing 5 eggs each. C individuals have a probability of 0.9 of having 0 offspring hatch and a probability of 0.1 of having 10. D individuals have a probability of $(0.9)^2$ of having 0 offspring, a probability of 2(0.9)(0.1) of having 5, and a probability of $(0.1)^2$ of having 10. The expected number of offspring is the same (= 1), but the C strategy of putting all eggs in a single nest has the higher variance in outcomes. Again for a simple example, suppose the population starts with one individual of each type. In the next generation, there will be either 0 C individuals or 10, and there will be either 0, 5, or 10 D individuals.

[12] The variance of a quantitative variable is its average squared deviation from the mean. Trait B's variance in fitness is $\left(\frac{1}{2}\right)\left[(2-1)^2+(2-3)^2\right]=1$.

[13] Here "expected frequencies" means the mathematical expectation.

Table 2.1 *The probability of six population configurations in the offspring generation and the [bracketed] proportion of C individuals in each*

		Number of D individuals		
		0	5	10
Number of C individuals	0	$0.9(0.9)^2$ [-]	$0.9(2)(0.1)(0.9)$ [0/5]	$0.9(0.1)^2$ [0/10]
	10	$0.1(0.9)^2$ [10/10]	$0.1(2)(0.1)(0.9)$ [10/15]	$0.1(0.1)^2$ [10/20]

Table 2.1 describes the probabilities of the six possible configurations of the offspring generation and the fraction of C individuals in each. The upshot is that the expected frequency of C in the offspring generation is

$$= \frac{(1)(0.1)(0.9)^2 + \left(\frac{2}{3}\right)(2)(0.1)^2(0.9) + \left(\frac{1}{2}\right)(0.1)^3}{1-(0.9)^3} = 0.345,$$

so the expected frequency of D is 0.655. The trait with the higher variance in outcomes declines in frequency (in expectation). I hope this example reminds you of an aphorism – *Don't put all your eggs in one basket.* Just as variance in outcomes matters when people decide how to invest time, money, and effort in their projects, so variance in offspring number matters to fitness in evolutionary biology. The phrase "bet hedging" applies to both.[14]

Evolutionary biologists use the following rule of thumb: Fitter traits increase in frequency in a population and less fit traits decline when selection controls the evolutionary process.[15] This means that a proper

[14] **Question:** How can a bird that puts its eggs in two nests be *reducing* variance, as compared to a bird that puts all its eggs in one, since the first bird has *more nests* than the second?

[15] The meaning of this rule is clear when there are just two traits to consider, but what if there are three or more? For example, consider three mutually exclusive traits A, B, and C, where A is fitter than B, and B is fitter C. The rule tells you that A will increase in frequency and C will decline, but what happens to B? The answer is that a trait increases in frequency (when selection is in control) precisely when the trait's fitness is greater than the average fitness of all the traits. More precisely, if A, B, and C have frequencies in the population of f_a, f_b, and f_c respectively (where $f_a + f_b + f_c = 1$), and the traits have fitnesses w_a, w_b, and w_c, then a trait increases in frequency precisely when its fitness exceeds $w_af_a + w_bf_b + w_cf_c$.

definition of fitness must take account of factors that go beyond an organism's viability and its expected number of offspring if it reaches adulthood. Variance matters too. Notice how the theoretical role played by fitness in modeling selection processes is relevant to defining what fitness is (Sober 2001a).

So far I've described three "components of fitness" – an organism's viability (= its probability of surviving from conception to reproductive age), its fertility (= its expected number of offspring if it reaches maturity), and its variance in offspring number. This list of relevant factors is not complete; if the expected number and the variance matter, so do skew and kurtosis (Beatty and Finsen 1989); these are further properties that a probability distribution has. I have not described how these different considerations should be assembled into an equation that defines fitness as a definite mathematical function of these and other relevant factors. That's an interesting question, but I will not pursue it here; see Frank and Slatkin (1990), Pence and Ramsey (2013), Hansen (2017), and Maxwell (2024 unpublished) for discussion. None of these points undermines the idea that an organism's fitness is an ability the organism has.

2.4 Estimating Fitnesses

The bias of a coin and the viability of an organism are both abilities that can be defined by using the concept of probability, but there's an important difference between the two. Coins can be tossed repeatedly, but organisms taste of life but once. Just as tossing a coin once does not allow you to obtain a good estimate of its probability of landing heads, you also can't estimate the probability a single organism has of surviving to reproductive age on the basis of whether it does so. True, if the organism in question were cloned a thousand times and the clones were reared in identical environments, the result of that experiment would provide ample evidence for estimating the organism's probability of surviving. However, this possibility does you no good if the organisms you're actually looking at have different genotypes and phenotypes, live in different environments, and there's no way you can carry out the nice experiment just described.

My pessimism about estimating the viabilities of single organisms has an exception. If you know that the organism has a trait that will kill it before it reaches reproductive age no matter what other traits it has, you can conclude that the organism has a viability fitness of zero. My claim therefore is this: The viabilities of single organisms *typically* cannot be estimated.

Fertility is a somewhat different story. An organism has just one life to live, but organisms often have repeated opportunities to reproduce, and in each bout of reproduction, they might produce different numbers of offspring; such organisms are said to be *iteroparous*. Perhaps the lifetime reproductive success of these organisms permits one to estimate their expected number of offspring.[16] However, this fact about their fertility doesn't save the day for estimating an organism's overall fitness. This quantity depends on both viability and fertility, which means that an organism's overall fitness typically can't be estimated, since its viability typically can't be estimated.[17]

If the overall fitness of an individual organism usually can't be estimated, the same point holds for the fitness of a special type of trait. Think of *all* the genetic and phenotypic traits (T_1, T_2, ..., T_n) that a single organism has, and now consider the conjunction of these traits (T_1&T_2&...&T_n). That conjunction describes what I'll call the organism's *total trait complex*. There is a connection between an organism's fitness and the fitness of the total trait complex that the organism has:

(CONNECT-1) The overall fitness of individual i in environment e = the fitness of the total trait complex that i has in e.

This principle entails something that I think is obvious – that two individuals that have exactly the same total trait complex and live in exactly the same environment must have the same fitness.[18] Philosophers put this point by saying that an organism's fitness *supervenes* on the totality of traits that it and its environment have. Organisms can't differ in fitness unless they have different traits or live in different environments.[19] My argument

[16] The inferential situation is more dismal for organisms that reproduce just once; they are *semelparous*. Pacific salmon swim upstream to the place where they were born, reproduce, and die.

[17] Pence and Ramsey (2015, p. 1089) disagree, arguing that the fitnesses of single organisms are knowable, but what they in fact point out is just that a single organism's survival and degree of reproductive success are knowable.

[18] In similar fashion, two coins that are identical in their physical properties and that are tossed in exactly the same way must have the same probability of landing heads.

[19] Supervenience is a determination relation, but the relationship of an organism's total trait complex and its environment at time t to the organism's fitness at time t is

that a single organism's fitness is typically unknowable entails that the fitness of its total trait complex is too.

The problem faced here is different from the tautology problem. The tautology problem has to do with *defining* fitness; its solution is to recognize that fitness is an ability. The present problem concerns *estimating* fitness values, once you grant that fitness is an ability. Evolutionary biology deploys a neat solution to the estimation problem. It changes the subject. Instead of trying to estimate the fitnesses of single organisms (and the fitnesses of total trait complexes), it sets its sight on estimating the fitnesses of traits that are found in multiple organisms. I'll call these traits "single traits" to distinguish them from total trait complexes; the latter are large conjunctions, while the former are the conjuncts that appear in multiple conjunctions. Consider the single trait of *running fast*. The fast zebras in a population at a given time are different from each other in numerous respects that affect their fitness, but you can estimate the average viability of those fast zebras by observing the frequency with which they survive from birth to reproductive age. Estimating the viability of a single trait is like taking a thousand differently shaped coins and tossing each just once. The observed frequency of heads doesn't tell you what *each* coin's probability of landing heads is, but it does allow you to estimate the *average* probability of landing heads in that heterogeneous hundred-coin ensemble.

The shift from thinking about the fitnesses of single organisms to thinking about the fitnesses of single traits fits nicely with the overall objectives of evolutionary biology. Evolution by natural selection is a multi-generational process. Discussions of evolution by natural selection focus on a trait's change in frequency through multiple generations. However, every sexual organism in the history of a species has a unique total trait complex.[20] It is *single* traits (and some conjunctions of single traits) that are exemplified in multiple generations in a population and change their frequencies.

Evolutionary biologists have developed numerous models that characterize which traits are fitter than which others in this or that circumstance.

not causal if cause must precede effect. The relationship of an organism's fitness to its other traits resembles the relationship of a coin's probability at time t of landing heads if it is tossed to the physical characteristics of the coin and its tossing environment at time t.

[20] What about so-called *identical twins*? It is better to call them *monozygotic*. Their environments aren't identical, and mutations occur in one twin that do not occur in the other.

My brief discussion of sex ratio evolution provided a glimpse of what those theories look like, as does Gillespie's example of not putting all your eggs in one basket. I'll discuss other such models in Chapters 3 and 8. Do such models solve the problem of figuring out what the fitnesses are of single organisms? No, these models are about single *traits*, not single *organisms*. For example, when there is selection for the sex ratio strategy of producing all daughters, this doesn't mean that the parents who deploy that strategy must have the same fitness. That's a good thing, since those parents inevitably differ from each other in numerous respects that affect their fitnesses.

2.5 Type and Token

I hope my contrast between the fitness of a *single* organism and the fitness of a *single* trait is clear as it stands, but the double use of the word "single" may be a stumbling block. This impediment is easily removed by using the standard philosophical distinction between *token* and *type*. Consider the ambiguity that is involved in saying that Jack and Jill own the same book. This might mean that there are two token books that are of the same type, one owned by Jack and the other by Jill, but it also can mean that there is a single token book that they co-own. The view I defended in the previous section is that the fitnesses of *token* organisms are mostly unknowable, but the same is often not true of the fitnesses of *types* of organism.

The type/token distinction applies to other topics in the philosophy of evolutionary biology. The claim of universal common ancestor (Section 1.1) is a claim about a token ancestor; this goes beyond the claim that present living things trace back to multiple token ancestors that share some of their traits. The biological terms "genotype" and "phenotype" accord with philosophical usage, as they refer to types, not to tokens. However, biologists sometimes talk about gene tokens in models of random genetic drift and in models of coalescence, as I'll explain in Chapters 5 and 9.

Fallacies sometimes arise by confusing type and token. Williams (1966, pp. 22–24) argues that genes are the "units of selection" (a concept I'll explore in the next chapter) because the same gene can exist in a long stretch of generations; in contrast, he says that organisms and groups can't be units of selection because they exist for much shorter stretches of time. Dawkins (1982) repeats this argument. What is true is that a gene *type* can be exemplified in indefinitely many generations, but the same is true of the *properties* of organisms and groups. On the other hand, gene *tokens* are at least as evanescent as token organisms and token groups.

2.6 Trait Fitnesses Are Averages

Although I've taken pains to distinguish the fitness of a token organism from the fitness of a trait, I noted a special context in which the two have the same value. As stated in the CONNECT-1 proposition, the fitness of a token organism in a given environment is identical with the fitness of that organism's total trait complex in that environment. An organism's total trait complex needs to be distinguished from the numerous singleton traits it has, but CONNECT-1 can be used to derive a second proposition that describes how the fitness of a trait is connected to the fitnesses of token organisms (Sober 2013):

> (CONNECT-2) The fitness of trait T in population p at time t = the average fitness of the token individuals that have T in p at t.

To see why CONNECT-2 follows from CONNECT-1, let's consider a simple example in which x, y, and z are the only individuals in the population that have trait T at a given time, and those individuals have total trait complexes X, Y, and Z, respectively. For simplicity, let's restrict our attention to viability fitnesses. To begin, note what the viability fitness of a trait is:

> The viability fitness of trait T = Pr(individual i survives to reproductive age | i has T)

Notice that the conditional probability[21] in this statement involves conditionalizing just on the fact that i has T; it says nothing more about i's total trait complex. Since there are only three individuals that have trait T in this example, you can describe the viability fitness of trait T as follows:

> Pr(individual i survives to reproductive age | i has T) =
> Pr(individual i survives to reproductive age | i has X)Pr(i has X | i has T) +
> Pr(individual i survives to reproductive age | i has Y)Pr(i has Y | i has T) +
> Pr(individual i survives to reproductive age | i has Z)Pr(i has Z | i has T)

[21] "Pr(B | A)" should be read as the probability that B would have if A were true ("the probability of B given A").

There are three addends on the right-hand side of this equation, and each addend has two product terms. In each addend, the first product term is the fitness of a total trait complex, which is identical with the fitness of a token organism (organism x or y or z, as the case may be). The second product term describes a token individual's probability of having a given total trait complex, given that it has singleton trait T. Using the actual frequency interpretation of probability, which is what biologists do in this context, each of those probabilities equals $\frac{1}{3}$.[22]

CONNECT-2 has an important consequence. Two traits have identical fitnesses in a population at a given time if they attach to exactly the same organisms in that population at that time. Using terminology from logic, the claim is that *coextensive traits are equally fit.*[23]

When two coextensive phenotypes are present in some but not all the organisms in a population, the question arises of why they are coextensive. Two possible answers are suggested by genetics. A given gene may cause two or more phenotypes. This is what it means for the gene to be *pleiotropic*, a topic I'll return to in Chapters 6 and 8. Another genetic source of coextensive phenotypes is gene linkage. If two genes are close together on the same chromosome and have different phenotypic effects, the pair of genes will probably be sent to offspring as a package, and so will the pair of phenotypes.

2.7 Fitness ≠ Advantageousness

Coextensive traits have identical fitnesses even if one of the traits is advantageous while the other is not. For example, if organisms with hemoglobin in their blood all have blood that is red, and organisms without hemoglobin do not have red blood, then the fitness of having hemoglobin is identical with the fitness of having red blood. It doesn't matter that hemoglobin helps individuals by transporting oxygen to their

[22] Pence and Ramsey (2015) argue that "organismic fitness ... lies at the conceptual basis of trait fitness," thinking that I have denied that claim. Their use of the word "basis" indicates that they hold that trait fitnesses are metaphysically grounded in the fitnesses of token organisms. I agree that the two concepts are definitionally connected.

[23] The extension of a trait includes all and only the token individuals that have that trait.

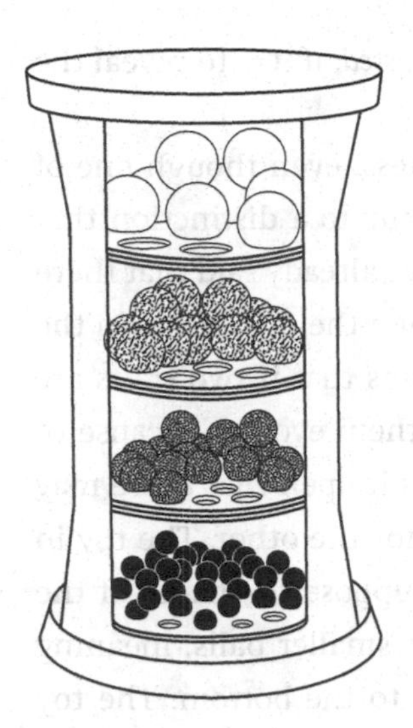

Figure 2.1 A selection toy.

tissues, while having red blood doesn't help organisms at all. If individuals with hemoglobin in their blood are fitter than individuals without, then individuals with red blood are fitter than individuals who don't have red blood.

I presented this point in my first book in philosophy of biology, *The Nature of Selection* (Sober 1984), by talking about the "selection toy" depicted in Figure 2.1. The toy has four "levels" defined by disks that contain holes. Bigger holes are at the top and smaller holes are at the bottom. If you invert the toy and shake it, all the balls will fall to the bottom. If you invert the toy again and shake it, the balls get sorted, with the smallest balls finding their way to the bottom, while larger balls do not. The trait of being small and the trait of being dark apply to exactly the same balls, so the probability a small ball has of getting all the way to the bottom is the same as the probability a dark ball has of getting there. However, it's the size of a ball that causes it to get to the bottom or to fail to do so; color makes no causal difference.

It is obvious from the construction of the toy that it is size not color that causally influences how balls get sorted. Unfortunately, causal facts are often not so transparent. When that is so, an *intervention* is helpful. If you are not convinced that it is size, not color, that does the causing, reach into the toy and use a paintbrush to break the correlation between size and color; when you do so, be careful not to change the disks and their holes. Your intervention must be *surgical*, not *ham-fisted*, if it is to reveal the causal facts (Woodward 2004).

The fact that two traits can have the same fitness, even though one of them is advantageous while the other is not, points to a distinction that needs to be drawn concerning natural selection. I've already said that there is selection in a population at a time precisely when the organisms in the population differ in fitness at that time. This means that if two traits are perfectly correlated in a population, and one of them evolves because of selection, the other does too. However, this leaves it open that there may have been selection for one of the traits but not for the other. The toy in Figure 2.1 can be thought of as a selection toy. Suppose the name of the game is getting to the bottom. The toy selects for smaller balls, meaning that a ball's being small is what causes it to get to the bottom. The toy thereby selects dark balls, but there's no selection for being dark. The color of a ball does not cause it to get to the bottom. So both of these statements are true: "there's selection *of* small balls" and "there's selection *of* dark balls." However, only one of the following statements is true: "there's selection *for* being small" and "there's selection *for* being dark." Likewise, there was selection for having hemoglobin, but no selection for having red blood; even so, having hemoglobin and having red blood were both selected. The evolution of a useless trait *piggybacks* on the evolution of an advantageous trait if the two are correlated.[24]

Although there can be selection of a trait without there being selection for that trait, it is plausible to maintain that where there is selection-of, there is also selection-for. If balls with different colors have different probabilities of making it to the bottom of the toy when the toy is shaken, there must be some property of those balls that is doing the causal work, even if it is not their color.

The *selection-of* versus *selection-for* distinction is needed to define an important concept in evolutionary biology – the concept of adaptation:

[24] Walsh (2000, 2004), Walsh, Lewens, and Ariew (2002), and Matthen and Ariew (2002) argue that selection never causes traits to evolve; what causes evolution is the births and deaths of individuals. Shapiro and Sober (2007) criticize their idea that selection is epiphenomenal. If selection-for means what I have claimed it to mean, epiphenomenalists are committed to saying that selection for a trait never in fact happens.

> Trait T is now an adaptation for doing X in a population if and only if trait T evolved in the lineage leading to that population because there was selection for trait T and the reason there was selection for T was that T caused individuals to do X.[25]

The concept of *adaptation* looks to the past; to say that a trait now is an adaptation is to say something about the trait's history.[26] In contrast, the concept of a trait's being *adaptive* looks to the future; it describes how having the trait will help organisms henceforth. The heart is an adaptation for pumping blood, not for making noise, though the noise-making of a parent's heart may now be useful for putting babies to sleep and for allowing physicians to diagnose heart problems. This is an example of a trait's now being adaptive without its being an adaptation. The reverse possibility is exemplified by traits that evolved because they *were* useful to the organism but fail to be useful *now*. An example is our taste for fats and sugar; these cravings served our ancestors well when food supplies were uncertain, but these same tastes now are maladaptive for individuals in affluent societies, as witnessed by the widespread health problems caused by obesity.

2.8 Improvement in Fitness

If the evolution of a population is controlled by natural selection, does it follow that the organisms at the end of the process will be fitter than the organisms at the beginning? The answer seems to be *yes*, because with selection in control, fitter traits increase in frequency while less fit traits decline. However, the answer in fact is *no*; this "optimistic" picture of natural selection is often mistaken. For one thing, the physical environment may change faster than the population is able to evolve. Consider the

[25] To say that a trait is an adaptation is elliptical; the trait must be an adaptation *for performing some task*.

[26] The definition of adaptation resembles Larry Wright's (1973) etiological account of what the concept of function means. For a very different account, see Robert Cummins (1975), who argues that the function of a trait now is to be understood in terms of what the trait does now. My view is that you don't need to choose between these two ideas; each is useful, though you need to say which you mean when you use the term. See Garson (2023) for discussion.

Frequency independent fitnesses

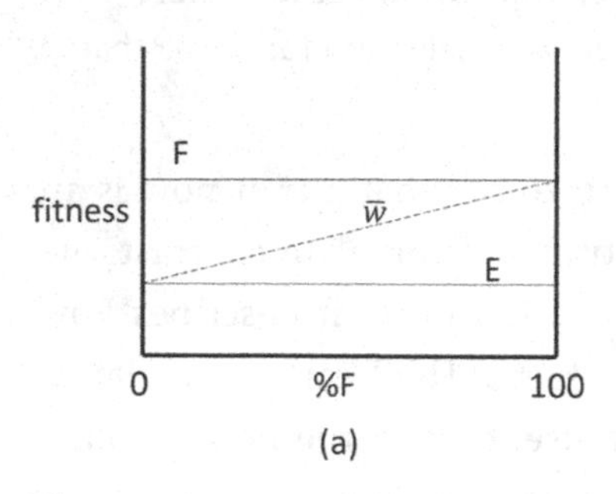

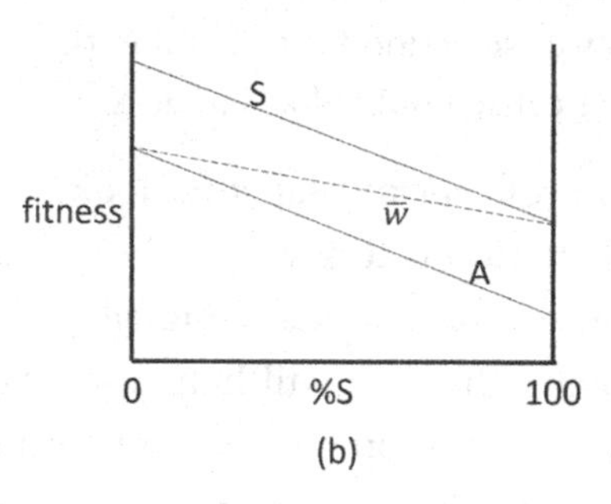

Figure 2.2 An optimistic and a pessimistic view of evolution by natural selection.

example of bear fur in the previous chapter; as the temperature drops, the bear species evolves thicker fur. Now suppose the temperature drops faster than the bear population evolves; this will happen if favorable mutations don't appear fast enough. The result is that bears at the end of the process are less well adapted to their *very* cold environment than the bears at the start were to their *somewhat* cold environment. This possibility is consonant with Leigh Van Valen's (1973) Red Queen hypothesis. In Lewis Carroll's *Through the Looking-Glass*, the Red Queen says that she needs to run as fast as she can just to remain in the same place. Van Valen's idea is that species need to evolve as fast as they can if they are to prevent their risk of extinction from increasing.

The optimistic picture of natural selection can go wrong even when a population's physical environment remains unchanged. When a population evolves, its trait frequencies change (by definition), and this shift can change the fitness values of traits. Figure 2.2 illustrates this point by contrasting frequency independent selection with frequency dependent selection. For the sake of a simple example, I assume that the organisms in the population are asexual and that each organism has one or the other of the two traits represented. Mutations aside, offspring have the same traits as their parents. If a population starts at 100 percent E in Figure 2.2a and an F mutation then occurs, what will happen? If natural selection controls the evolutionary process, F will increase in frequency and will eventually reach 100 percent because F is fitter than E. As F increases in frequency, the average fitness of the individuals in the population (represented by

$\bar{w}$, pronounced "w-bar") increases. This is the optimistic picture of selection improving fitness. For example, suppose that zebras are either fast or slow, and that fast zebras are always fitter than slow zebras because fast zebras are more able to evade lion attacks. In addition, let's suppose that lions decide which zebra to attack at random; they don't preferentially attack slow zebras. In this case, the fitnesses of the traits do not change as the population evolves from 100 percent Slow to 100 percent Fast. The fitnesses here are *frequency independent*.

Figure 2.2b tells a different story. Suppose the population is 100 percent A and then an S mutation occurs. If selection controls the evolutionary process, S will increase in frequency and will eventually reach fixation. As S increases in frequency, $\bar{w}$ declines. This is the pessimistic picture of selection reducing fitness. Figure 2.2b will be important in the next chapter, where I'll discuss the evolution of altruism (A) and selfishness (S). Selfish individuals are always fitter than altruists in the same group, but everyone in the group benefits the more altruists there are. As selfishness replaces altruism in the group, everyone's fitness declines. These relationships define what "selfishness" and "altruism" mean in evolutionary biology.[27,28] By definition, their fitnesses are *frequency dependent*.

Selection pressures on a population are often generated by the physical environment, as in the example of bears living in a climate that gets colder. The present discussion of frequency dependent selection shows that the population itself can have features that create selection pressures. I described an example of frequency dependent selection earlier, in the brief discussion of sex ratio evolution. A population experiences selection

[27] **Questions:** In the first edition of the *Origin*, Darwin (1859, p. 489) says that "as natural selection works solely by and for the good of each being, all corporeal and mental endowments will tend to progress towards perfection." Darwin (1959, p. 758) left this sentence unchanged in all subsequent editions. How does frequency dependent selection bear on Darwin's claim? And how is the type/token distinction (Section 2.5) relevant to evaluating his claim?

[28] R.A. Fisher, in his 1930 book The Genetical Theory of Natural Selection states that "the rate of increase in fitness of any organism at any time is equal to its genetic variance in fitness at that time." Fisher calls this proposition *the fundamental theorem of natural selection*. **Questions:** (i) How does the type/token distinction apply to Fisher's theorem? (ii) The concept of variance used in statistics represents the amount of variation in a population. Given that variance must be non-negative, is Fisher's theorem compatible with the example of frequency dependent selection depicted in Figure 2.2b?

pressures precisely when there is variation in fitness; the source of that variation may be outside the population, inside, or both at once.

2.9 The Groups Above, the Genes Below

In the selection processes described in this chapter, it is the individual organisms in a single population that differ in fitness. In the next chapter, the picture expands in two directions. I'll talk about fitness differences among groups of conspecific organisms in a single "meta-population" and also about fitness differences among the genes inside a single organism. This will open the door to examining two new types of selection process – group selection and intragenomic conflict.

3 Units of Selection

3.1 Darwin on Altruism and Group Selection

The honeybee's barbed stinger was the example I used in Section 1.5 to explain why Darwin included group selection in his theory. The barb is good for the nest but bad for the individual honeybee, which eviscerates itself when it stings an intruder to the nest. Here "good" and "bad" are defined by their fitness effects.

Another puzzle that drove Darwin to the hypothesis of group selection was the existence of sterile workers in several species of social insect. He says that "this is by far the most serious special difficulty which my theory has encountered" (Darwin 1859, p. 242). Sterile individuals have overall fitnesses of zero, so how can sterility evolve and be maintained by natural selection? Here is Darwin's answer:

> How the workers have been rendered sterile is a difficulty; but not much greater than that of any other striking modification of structure; for it can be shown that some insects and other articulate animals in a state of nature occasionally become sterile; and if such insects had been social, and it had been profitable to the community that a number should have been annually born capable of work, but incapable of procreation, I can see no very great difficulty in this being effected by natural selection (Darwin 1859, p. 236).

Groups that contain both sterile and fertile individuals outcompete groups that contain only fertile individuals if sterile individuals provide sufficient fitness benefits to their fertile nestmates. Rather than focusing on the zero fitness of sterile workers, think about two parental strategies. The All-F strategy is to produce fertile offspring; the Mixer strategy is to produce a mix of sterile offspring and fertile offspring, where the sterile offspring help everyone in the nest, regardless of whether their parent was an All-F or a Mixer.

Darwin also applies the idea of group selection to human evolution in his 1871 book, *The Descent of Man and Selection in Relation to Sex*. He does so to explain a striking fact about human morality. Darwin first poses the problem:

> It is extremely doubtful whether the offspring of the more sympathetic and benevolent parents, or of those which were the most faithful to their comrades, would be reared in greater number than the children of selfish and treacherous parents of the same tribe. He who was ready to sacrifice his life, as many a savage has been, rather than betray his comrades, would often leave no offspring to inherit his noble nature. The bravest men, who were always willing to come to the front in war, and who freely risked their lives for others, would on an average perish in larger number than other men (Darwin 1871, p. 163).

Here is Darwin's solution:

> It must not be forgotten that although a high standard of morality gives but a slight or no advantage to each individual man and his children over the other men of the same tribe, yet that an increase in the number of well-endowed men and an advancement in the standard of morality will certainly give an immense advantage to one tribe over another. A tribe including many members who, from possessing in a high degree the spirit of patriotism, fidelity, obedience, courage, and sympathy, were always ready to aid one another, and to sacrifice themselves for the common good, would be victorious over most other tribes; and this would be natural selection. At all times throughout the world tribes have supplanted other tribes; and as morality is one important element in their success, the standard of morality and the number of well-endowed men will thus everywhere tend to rise and increase (p. 166).

Darwin wondered why human societies almost always contain moral norms that require group members to sometimes sacrifice their own welfare for the sake of other group members. Societies of course differ considerably in the details of the norms they deploy, and people don't always conform to their society's norms. However, people often obey those norms, at cost (large or small) to themselves, where the cost involves reduced prospects of surviving or reproducing. So why do norms of self-sacrifice persist? Darwin thought that groups of human beings have long been in competition with each other, sometimes through warfare (like the two dogs fighting over a piece of meat described in the previous chapter), and sometimes through their differential ability to avoid extinction and to found new groups (like the two plants at the edge of the desert). The result is that norms of self-sacrifice evolved by a process of group selection. Self-sacrifice in human beings resembles the barbed stinger of honeybees and the sterile workers in species of social insect, despite the enormous differences between the traits and the species that have them.

Evolutionary biologists now use the term "altruistic" to describe individuals who reduce their own fitness while augmenting the fitness of others.[1] The term "selfish" is used to denote the opposite pattern; selfish individuals do not reduce their own fitness nor do they increase the fitness of others. "Altruism" and "selfishness" are terms in our everyday speech, but evolutionary biologists give them a special twist. According to their definitions, you don't need to have a mind to be altruistic or selfish.[2] The terms refer only to the fitness effects of traits. The definitions of altruism and selfishness entail that selfish individual are fitter than altruists in the same group, whereas groups in which there are more altruists are fitter than groups in which there are fewer.[3]

The view I'll explain and defend in this chapter is called multi-level selection (MLS) theory. It's the idea that selection can and does occur at multiple levels of organization. In group selection, groups compete with each other. In individual selection, it's individuals in the same group that compete. Group selection promotes the evolution of altruism while individual selection promotes the evolution of selfishness. When both selection processes occur in a meta-population of groups, they push in opposite directions. Whether altruism increases in frequency or declines depends on which force is stronger.[4]

[1] The word "altruism" was invented by the philosopher August Comte in the 1850s to name a concern for the welfare of others. For discussion of its impact on Victorian culture, see Dixon (2008).

[2] For discussion of *psychological* egoism and altruism, see Part 2 of Sober and Wilson (1998).

[3] This claim needs to be qualified in connection with hyper-altruism, which I discuss in Section 3.4. I thank Shimin Zhao for this point.

[4] I find it useful to use the word "force" when describing how group and individual selection affect the evolution of altruism. Forces are causes, but the word "force" means something more. Forces are *magnitudes*, but causal statements often don't involve quantitative variables. If the movie's being sad caused Joe to weep, I would not describe this by saying that there was a force of sadness that impinged on Joe. In addition, forces

Although Darwin introduced the idea of group selection because he was thinking about altruism, and altruism was the central focus when biologists talked about group selection in the twentieth century, the idea of group selection is not limited to that example, as I'll explain in Section 3.9. This means we need to have definitions of group and individual selection that don't mention altruism and selfishness. MLS theory defines individual selection as fitness variation within a group, and group selection as variation in fitness among groups in a meta-population.[5]

So far I've mentioned two levels of organization at which selection occurs, but there is a third, which Darwin did not anticipate. *Intragenomic conflict* occurs when genes in the same organism compete. I discuss this in Section 3.10.

3.2 Simpson's Paradox

If there are several groups in a meta-population and each contains both selfish and altruistic individuals, then altruists are less fit than selfish individuals in each of those groups. This is true by definition. Using the rule of thumb described in the previous chapter – when selection controls the evolutionary process, fitter traits increase in frequency and less fit traits decline – we seem driven to the conclusion that selfishness must increase in frequency in each group and altruism must decline. From this, it seems to follow that altruism cannot evolve by natural selection. And if it can't evolve by *natural* selection, it can't evolve by *group* selection either, since group selection is a kind of natural selection. Did Darwin goof when he invented the idea of group selection?

have *direction*, and component forces combine to form a resultant. For those who find my use of the word "force" objectionable, I suggest that they read it to mean *cause*.

A properly formulated conception of force requires one to describe what happens when the force is zero. Newton's law of inertia says that an object will remain at rest or in uniform motion if no forces impinge. In population genetics, the Hardy–Weinberg law is properly viewed as a zero-force law. However, there should be a non-genetic zero-force law for evolution, since the genetic system itself evolved. McShea and Brandon (2010) attempt to describe this more general law; see Barrett et al. (2012) for discussion.

[5] **Question:** In Section 2.3, I discussed sex ratio in the context of a single group, explaining how there is within-group (= individual) selection for producing offspring of the minority sex, the effect of which is the evolution of an even sex ratio. What sex ratio would evolve if group selection controlled the evolution of sex ratio?

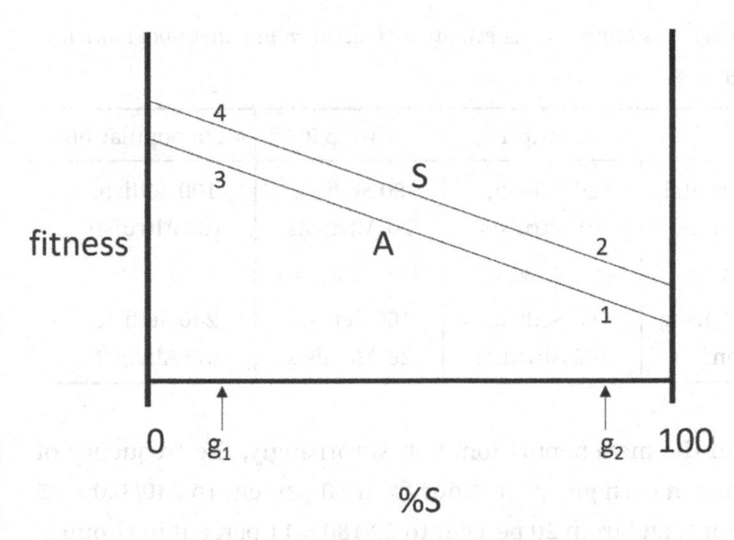

Figure 3.1 An example of group and individual selection when there are two groups.

There is a fallacy in the argument just described. Even though altruists are by definition less fit than selfish individuals *in the same group*, it does not follow that altruists are on average less fit than selfish individuals *in the meta-population as a whole*. The rule of thumb indicates that altruism must decline in each group, but (surprisingly) it does not follow that altruism must decline in the meta-population. This may sound crazy; if a trait is shrinking in frequency in each state in the United States, doesn't it follow that the trait is shrinking in frequency in the whole United States?

A simple example explains why what seems crazy isn't crazy at all. Consider Figure 3.1, which adds some details to Figure 2.2b from the previous chapter. The additions are the mentions of two groups (g_1 and g_2) that differ in their frequencies of selfishness. For g_1, the frequency is 20 percent; for g_2, it's 80 percent. Let's assume that these individuals reproduce asexually and that altruistic parents always have altruistic offspring and that selfish parents always have selfish offspring. What frequencies will altruism and selfishness have in the offspring generation?

Table 3.1 describes the upshot. For the moment, think of fitness as the *actual* number of offspring a parent will have, given its own phenotype and the kind of group it is in. There are equal numbers of altruists and selfish individuals in the parental generation, but there are 260 altruists and 240 selfish individuals in the offspring generation. The frequency of altruism

Table 3.1 *Hypothetical example of the evolution of altruism in a meta-population containing two groups*

	Group 1	Group 2	Meta-population
Number in parental generation	20 Selfish, 80 Altruists	80 Selfish, 20 Altruists	100 Selfish, 100 Altruists
Fitnesses	$w_S = 4$, $w_A = 3$	$w_S = 2$, $w_A = 1$	
Number in offspring generation	80 Selfish, 240 Altruists	160 Selfish, 20 Altruists	240 Selfish, 260 Altruists

has increased in the meta-population, but, surprisingly, the frequency of altruism declined in each group. It drops from 80 percent to 240/320 = 75 percent in Group 1, and from 20 percent to 20/180 = 11 percent in Group 2.

In this example, selfish individuals are fitter than altruists in each group (since 4 > 3 and 2 > 1), but the average fitnesses of the two traits in the meta-population are

$$\text{Fitness of A} = (80\%)(3) + (20\%)(1) = 2.6$$
$$\text{Fitness of S} = (20\%)(4) + (80\%)(2) = 2.44.$$

The average fitness of altruism is *greater* than the average fitness of selfishness. These average fitnesses belong in the blank cell of Table 3.1.[6]

Here's the fallacious argument I've been discussing:

In any group that contains both altruists and selfish individuals, the selfish individuals are fitter.

When selection controls the evolutionary process, fitter traits increase in frequency and less fit traits decline.

Therefore, when selection controls the evolutionary process, altruism cannot increase in frequency.

The two statements above the line are the argument's premises; the statement below is the conclusion. The premises are true, but in the example just described, the conclusion is false. The argument therefore fails to be deductively valid. This defect can be repaired if the conclusion of the argument is replaced. What the argument should have as its conclusion is that

[6] **Exercise:** Reformulate this example so that fitnesses describe *expected* numbers of offspring, not *actual* numbers.

Table 3.2 *Hypothetical example of Simpson's paradox in university admissions*

	Department 1	Department 2	University
Number applying	90 Women, 10 Men	10 Women, 90 Men	100 Women, 100 Men
Percentage admitted	20 percent	40 percent	
Number admitted	18 Women, 2 Men	4 Women, 36 Men	22 Women, 38 Men

when selection controls the evolutionary process, altruism cannot increase in frequency *within groups*. However, it can increase *in the meta-population*. The difference between what happens in the *parts* and what happens in the *whole* is key.

This reasoning about wholes and parts has a much broader application; it has nothing essentially to do with evolution or altruism (Sober 1984). It now is called "Simpson's paradox," named for the statistician E. H. Simpson (1951), not for George Gaylord Simpson, the biologist I cited in the previous chapter in connection with the idea that selection is "opportunistic." Here's an example of Simpson's paradox that Nancy Cartwright (1979) described. The University of California at Berkeley found evidence that there was sex discrimination in admission to graduate school. It was observed that

(1) A smaller percentage of women were admitted to graduate school than men.

However, when the university looked more closely at the evidence, they also found that:

(2) In each department, the percentage of women admitted was the same as the percentage of men.

To see how (1) and (2) can both be true, consider the example depicted in Table 3.2. There are two departments (D_1 and D_2) in the university and 100 women and 100 men apply to graduate school. In accordance with (1), suppose that 22 women and 38 men are admitted. Given this, how can proposition (2) be true? The answer is to be found in a further fact – that the admissions rates for men and women in each department are the same, but D_1 admits 20 percent of its applicants while D_2 admits 40 percent.

Here is a fallacious argument: "Since each department admits the same percentage of women as men, the university as a whole must admit the same percentage of women as men." This line of reasoning would be correct if there were no *correlation* between an individual's gender and the department to which the individual applies. However, in the present example, there is: Pr(applying to D_1 | female) > Pr(applying to D_1 | male) and Pr(applying to D_2 | male) > Pr(applying to D_2 | female).[7,8]

3.3 The Problem of Origination

In the two-group example just discussed, the meta-population begins with altruism having a frequency of 50 percent. The example shows how altruism can increase in frequency given that starting point. This can't be the whole story about altruism's evolution, however, since it fails to explain how selection can increase the frequency of altruism after the trait is first introduced into the population by a mutation. This is called *the problem of origination*. A lone mutant altruist has a lower fitness than the average selfish individual in the meta-population, by definition. This fact is represented in Figure 3.1.

George Williams and Doris Williams (1957) showed how the problem of origination can be solved. Their solution begins with a distinction between genotypes and phenotypes. Suppose that

- Altruism is caused by a single dominant gene A, so AA and Aa individuals are altruistic.
- Selfishness is caused by a single recessive gene a, so aa individuals are selfish.

[7] B is positively correlated with A precisely when Pr(B | A) > Pr(B | notA). This definition describes the correlation of propositions (or, equivalent, of dichotomous variables). Correlation is a symmetrical relation; if A is positively correlated with B, then B is positively correlated with A. Don't confuse the positive correlation of A and B with this inequality: Pr(B | A) > Pr(notB | A). This last inequality just means that Pr(B | A) > ½. **Exercise:** Show that this last inequality is neither necessary nor sufficient for B and A to be positively correlated.

[8] **Exercise:** Gildenhuys (2003) criticizes the claim that the evolution of altruism involves Simpson's paradox and argues that the evolution of altruism does not involve group selection. Assess his critique.

Table 3.3 *An example of Williams and Williams's (1957) solution to the problem of origination*

	Group X		Each of the many other groups
Parental pair founding a nest	Aa × aa		aa × aa
Offspring	50 percent Aa	50 percent aa	100 percent aa
Fitnesses of offspring	2	3	1.5

Suppose each group in a meta-population is founded by a single fertilized female laying her eggs in a nest; when the eggs in each nest hatch, nest-mates are siblings. Nest-mates interact with each other as they grow up (with altruists conferring fitness benefits on their siblings), after which they disperse from the nest, mate, and then fertilized females lay their eggs in new nests, thus creating the next generation of sibling groups. Initially, the meta-population is 100 percent selfish (so each individual is aa). Then a mutation occurs that introduces the A gene into the population, so now there is one individual (suppose it's female) that is Aa. This Aa individual then mates with an aa male and she lays her eggs in a nest. The eggs hatch and the nest is 50 percent altruistic (Aa) and 50 percent selfish (aa). Call this special group "group X." All the other nests in the population are 100 percent selfish.

Using the numbers in Figure 3.1, I've inserted numerical values for the fitnesses of the offspring in Table 3.3. Selfish individuals in group X are fitter than selfish individuals in the other groups, because the former live with altruists. Notice also that the altruists in group X are less fit than their selfish siblings (as the definitions of altruism and selfishness require). However, altruists are on average fitter than selfish individuals in the entire meta-population, since there are many aa individuals with fitnesses of 1.5 and only a few that have a fitness of 3.

This simple argument from Williams and Williams is what scientists call a "proof of concept," meaning that it shows that it's *possible* for natural selection to cause altruism to increase in frequency once an altruistic mutation has occurred. However, don't be surprised if the real genetics of altruism and selfishness in a species is often more complicated than the simple story just told.

Having addressed the problem of origination, I now want to consider a problem at the other end – if a meta-population somehow becomes 100 percent altruistic, is that a stable configuration? The answer is *no* – a mutant or migrant selfish individual will be at a reproductive advantage when it takes up residence with a bunch of altruists, and altruism will then move from 100 percent to something less. This suggests that when altruism evolves, the result should be a polymorphic equilibrium, meaning a configuration in which altruism and selfishness are both present.

This is a good place to comment on a more general topic – the assumption of *genetic determinism*. In the Williams and Williams model, genotypes determine phenotypes. This is almost always false. Almost all phenotypes are affected by both genes and environment. Given this, how can their model be a proof of concept? The answer to this question depends on what would happen if the assumption of genetic determinism were relaxed. If altruism and selfishness are probabilistically influenced by both genes and environment, will altruism be able to increase in frequency after the A mutation occurs? If the answer is *yes*, then the assumption of genetic determinism is in this case a *harmless idealization*.[9]

3.4 The Evolution of Altruism and Selfishness in Groups of Size 2

In Section 3.2, I discussed the bearing of Simpson's paradox on the evolution of altruism by giving a specific numerical example. I now want to describe a general criterion for when altruism will evolve in the context of a simple example in which the meta-population is composed of numerous groups of size two (Sober 1992, 1993). This means that there are three types of group – a group can have two altruists, two selfish individuals, or one of each.

Consider Table 3.4, which shows how your fitness depends on whether you are altruistic or selfish, and also on whether your partner (i.e., the other individual in your group) is altruistic or selfish. This model is *additive*.

[9] As mentioned in Section 2.2, the breeder's equation says that the response to selection equals the heritability times the strength of selection. Heritability is a quantitative variable whose values fall between 0 and 1. This shows that genetic determinism isn't essential for selection to cause a trait to evolve.

Table 3.4 *Your fitness depends on your phenotype and on the phenotype of your partner*

		Your partner is	
		Altruistic	Selfish
You are	Altruistic	$x+b-c$	$x-c$
	Selfish	$x+b$	x

If you shift from selfish to altruistic, you pay cost c regardless of whether your partner is altruistic, and if your partner shifts from selfish to altruistic, you gain benefit b regardless of whether you are altruistic.[10] Altruists have different fitnesses, depending on who their partners are; selfish individuals also differ in fitness, and for the same reason. Despite this variability, the *average* fitnesses of altruism and selfishness in the meta-population can be described in equations.

If you're an altruist, you have a probability p that your partner is an altruist and a probability $(1 - p)$ that your partner is selfish. That is, there are two conditional probabilities and they sum to 1:

$$p = \Pr(\text{your partner is altruistic} \mid \text{you are altruistic})$$
$$(1-p) = \Pr(\text{your partner is selfish} \mid \text{you are altruistic})$$

This means that your fitness, if you are an altruist, is

$$\text{Fitness of } A = p(x+b-c)+(1-p)(x-c)$$

Similarly, if you are selfish, you have a probability q that your partner is altruistic and a probability $(1 - q)$ that your partner is selfish. Here again, q and $(1 - q)$ are conditional probabilities that sum to 1:

$$q = \Pr(\text{your partner is altruistic} \mid \text{you are selfish})$$
$$(1-q) = \Pr(\text{your partner is selfish} \mid \text{you are selfish})$$

Thus, your fitness, if you are selfish, is

$$\text{Fitness of } S = q(x+b)+(1-q)(x)$$

[10] Non-additivity can be represented by making the model more complicated, as I'll explain.

It follows that

$$A \text{ is fitter than } S \text{ if and only if } p(x+b-c)+(1-p)(x-c) > q(x+b)+(1-q)x.$$

This simplifies to

(D) A is fitter than S if and only if $(p-q) > c/b$.[11]

Here $(p - q)$ is the correlation between your being altruistic and your partner's being altruistic. If $(p - q) = 0$, the probability of your partner's being an altruist is the same regardless of whether you are altruistic or selfish. If $(p - q) > 0$, then your chance of partnering with an altruist is greater if you are an altruist than it would be if you were selfish. I call this criterion "D" because it uses the concept of "Darwinian fitness" that I've used in this chapter and the ones before. That does not mean that Darwin wrote out the algebra.

The D criterion for the evolution of altruism in a meta-population when groups are of size two has two interesting implications. First, if $c > b$, altruism can't be fitter than selfishness. This is because p and q are probabilities, so $(p - q)$ can't be greater than 1. I'll use the term "hyper-altruism" to denote altruistic acts in which the cost to the altruist is greater than the benefit to the recipient; for example, consider the behavior of sacrificing one's life in order to make someone smile. According to the model, natural selection can't cause hyper-altruism to evolve. That doesn't mean that hyper-altruism can't exist; rather, the point is that some other causal process is needed to bring it into existence. The second implication is that if $(p - q) \leq 0$ and $c/b > 0$, then altruism can't be fitter than selfishness.

It is important to recognize that these two implications depend on the model's additivity. To see why, let's add an interaction term (i) to the upper-left cell in Table 3.4, so that an altruist interacting with another altruist has a fitness of $x + b - c + i$. The result is a new criterion:

$$A \text{ is fitter than } S \text{ if and only if } (p-q)b + pi > c$$

Now the evolution of altruism does not depend on $(p - q) > 0$, though that certainly helps.[12]

[11] **Exercise:** Prove that proposition D follows from the equations that describe the fitnesses of altruism and selfishness.

[12] **Question:** How do interactive fitnesses affect the possibility of hyper-altruism's evolving by natural selection?

I've said nothing about group selection in this section, but it's in the additive and the interactive models nonetheless. As mentioned, there are three kinds of group (*2A*, *1A* + *1S*, and *2S*). The fitness of a group is just the sum of the fitnesses of the individuals in the group, which means that the group fitnesses are:

Fitness of a *2A* group = $2(x + b - c + i)$
Fitness of a *1A* + *1S* group = $x + b + x - c$
Fitness of a *2S* group = $2x$

If $b > c$ and $i \geq 0$, then *2A* groups are fitter than *1A* + *1S* groups, which in turn are fitter than *2S* groups. Groups are fitter the more altruists they contain, so group selection favors altruism.

I've focused here on groups of size 2, but that was purely for convenience. The basic idea applies to groups of any size. Just as there are 3 kinds of group when individuals interact in pairs, there are N + 1 kinds of group when individuals interact in groups of size N. In such groups, the payoff matrix won't be a 2-by-2 table; it will be 2-by-(N – 1).

An interesting example of altruism and selfishness can be found in the evolution of infectious diseases. Viruses reproduce only when they commandeer the genetic machinery in the cells of host organisms. Each infected host houses a population of viruses. More virulent strains often reproduce more rapidly than less virulent strains within the same host. However, the more virulent the viruses in a host are, the lower is the host's life expectancy. Within each host, individual selection favors viruses that are more virulent over viruses that are less, but group selection will favor groups of viruses that are less virulent over groups that are more if new potential hosts are few and far between. Low virulence viruses are altruists, and high virulence viruses are selfish. When the Australian government introduced the myxoma virus in the 1950s to reduce the number of rabbits (which Europeans had brought to the continent), the rabbits evolved an enhanced resistance to the disease at the same time that the virus evolved a reduced virulence (Lewontin 1970b; Geoghegan and Holmes 2018).[13] The reduction in virulence was due to group selection.

[13] **Question:** How could scientists determine that both these events occurred if what they observed was just that rabbit mortality declined?

3.5 Fortuitous Group Benefits and the Definition of a Group

The fact that a trait is good for the group does not entail that it evolved *because* it was good for the group. Notice the contrast between the present tense "is" and the past tense "was" in the previous sentence. Present utility can be an imperfect guide to past evolutionary processes, a point I'll return to in Chapter 8. In particular, an individual selection process that occurs within a single group can cause traits to evolve that are good for the group. For example, consider Figure 2.2a from the previous chapter, which depicts the process of individual selection when fitnesses are frequency independent. To illustrate this idea, I told a simple story about fast and slow zebras in a single population. Fast is fitter than slow, and as Fast increases in frequency, the average fitness of the organisms in the population also increases. The population benefits from this transformation. A zebra herd has a lower probability of going extinct if everyone runs fast than it would have if everyone were to run slow. However, the trait of running fast did not evolve *because* running fast was good for the group. Rather, it evolved because running fast was good for individual zebras. The trait is not a *group adaptation*, since it did not evolve by the process of group selection. Rather, the group's reduced probability of going extinct is a *fortuitous group benefit*; the benefit to the group is a *side effect* of individual selection.[14]

This idea seems to raise a problem for the definitions I've given of group and individual selection. I said that individual selection occurs precisely when the individuals in a group differ in their fitnesses, and that group selection occurs precisely when the groups in a meta-population differ in their fitnesses. But now consider this example: suppose there are two herds of zebras – one is 100 percent Fast and the other is 100 percent Slow. There is no fitness variation within herds; all the variation is among herds. Does this mean that there is group selection here?

The answer is *no* if the fitnesses of Fast and Slow are frequency independent. In this example, there are no groups in the sense relevant to the units of selection issue, since an individual's fitness isn't *affected* by whether the other individuals in its herd are Fast or Slow; note that "affected" is a causal

[14] Okasha (2006) calls this a "cross-level byproduct."

verb. As far as running speed is concerned, individual selection is the only selection process that occurs in the single population comprised of the two herds. The framework I'm using classifies the trait of running fast as a purely individual adaptation.

Now for a complication: suppose a second selection process is under way at the same time in the two herds. Suppose the first herd is 20 percent selfish and the second is 80 percent. Individuals receive fitness donations only from altruists that are members of their own herd. This is the example depicted in Figure 3.1. The two herds are groups with respect to altruism and selfishness, but they aren't groups with respect to running speed. This means that the concept of group needs to be relativized to a trait, as follows:

> A set S of organisms whose members are $O_1, O_2, \ldots, O_n$ constitutes a group relative to trait T if and only if the fitness of each O_i in S depends on the frequency of T among the $n - 1$ other members of S, and O_i's fitness is not affected by whether O_j has T if O_j is not in S.[15]

As before, "depends" and "not affected by" should be understood causally. This is David Sloan Wilson's (1975) concept of a *trait group*. Note that the individuals in a trait group aren't required to reproduce with each other. For example, if lionesses cooperate when they hunt together, that makes them a trait group with respect to the trait of cooperative hunting.

Since group selection presupposes the existence of groups, it can't explain why group-living evolved. However, there's an interesting game-theory model that addresses this problem. It was inspired by a story told by the philosopher Jean-Jacques Rousseau in his 1755 *Discourse on Inequality*. Rousseau describes two hunters who each must decide whether to hunt stag or hunt hare without knowing what the other hunter will do. Each knows that a successful stag hunt requires the other hunter's participation whereas a successful hare hunt can be done solo. They also know that half a stag is worth more than a bag of hare. Rousseau concludes that the rational choice is to hunt stag.

To analyze Rousseau's idea, let's suppose that the individuals in a population each independently decide whether to hunt stag or hare, and then

[15] Here I'm talking about a dichotomous (two-state) trait; the definition can be generalized to trait variables that have more than two states.

Table 3.5 *Payoffs to row in the stag hunt game*

		Column Player	
		Hunt stag	Hunt hare
Row player	Hunt stag	$x+s$	x
	Hunt hare	$x+h$	$x+h$

randomly form up into pairs, meaning that the Row player's probability of pairing with a stag hunter is p, regardless of whether Row chooses to hunt stag or to hunt hare. The payoffs depicted in Table 3.5 show that Rousseau's conclusion does not follow unconditionally. This is because the expected utilities of the two traits are:

$$\text{Expected utility of hunting stag} = (x+s)p + x(1-p) = x + sp$$
$$\text{Expected utility of hunting hare} = (x+h)p + (x+h)(1-p) = x + h$$

Hunting stag has the higher expected utility precisely when $p > h/s$. By assumption, $s > h$. If $s >> h$, hunting stag can be the better choice even when the probability that the column player will hunt stag is small.[16] On the other hand, notice that if $p < h/s$, the rational choice is to hunt hare. The stag hunt game was originally posed as a problem about rational deliberation, and Rousseau lived long before von Neumann and Morgenstern (1944) invented deliberational game theory and Maynard Smith (1982) and others created evolutionary game theory. However, Rousseau's example shows how group living might first have evolved; just replace "expected utility" with "fitness."[17] The model shows how individuals who live in groups (and thereby affect each other's fitness) can do better than individuals living alone (Skyrms 2004).

Wilson's definition of group entails that a pair of hare hunters does not count as a group, since a hare hunter's fitness isn't affected by whether its partner hunts stag or hunts hare. There is just one kind of group in the stag

[16] The relation of pay-offs in the stag hunt game resembles the relation of pay-offs in Pascal's wager – his argument that it is prudent to believe in God even if you think that the probability that God exists is very small.

[17] **Exercise:** Construct a diagram for the stag hunt game in which the x-axis is the frequency of stag hunters in the meta-population and the y-axis is fitness. How does this diagram compare with the diagram for altruism and selfishness in Figure 2.2b?

hunt game; it is composed of two stag hunters. There is no group selection here, however, since these groups do not differ in fitness.[18]

3.6 Reciprocal Altruism

So far, I've discussed the evolution of altruism and selfishness in a meta-population of groups; for simplicity, I analyzed the problem by considering successive generations of a meta-population made of pairs of individuals. This evolutionary problem was not invented from scratch; rather, it was inspired by a problem in deliberational game theory called *the prisoners' dilemma*; there the behaviors are called "cooperate" and "defect" and the effects of those behaviors are represented in terms of utilities, where utilities represent whatever an agent wants and how much. The utilities in prisoners' dilemmas typically go far beyond surviving and reproducing.[19] Evolutionary game theory substitutes fitness for utility, and calls the behaviors "altruism" and "selfishness." Deliberational game theory provides normative advice to rational agents who want to maximize their expected utility. Evolutionary game theory doesn't provide advice to organisms, rather, it is a descriptive enterprise that aims at understanding which traits will evolve, in what circumstances, when selection controls the evolutionary process.

The prisoners' dilemma was formulated as a one-shot event. The players come together, decide what to do, then they act, and then they never see each other again. Will the game change if we move from one-shot to iterated, wherein players interact with each other repeatedly? Robert Trivers (1971) suggested that "reciprocal altruism" could help explain the evolution of cooperation without invoking group selection. This seemed promising since "you scratch my back, I'll scratch yours" can make sense even for agents who care only about themselves. Some years later, Robert Axelrod (1984) ran a game-theory tournament in which economists, game theorists, and biologists were asked to propose strategies that one individual could use in interacting several times with another; in

[18] A similar point applies to W.D. Hamilton's (1971) "selfish herd hypothesis," which says that group living evolves when prey organisms reduce their risk of predation by putting conspecifics between themselves and predators.

[19] See Poundstone (1993) for an historical introduction to deliberational game theory.

Table 3.6 *Payoffs to row on each round*

		Column player	
		Altruistic	Selfish
Row player	Altruistic	4	0
	Selfish	5	1

Table 3.7 *Total payoffs to row over fifteen rounds*

		Column player	
		TFT	ALL-S
Row player	TFT	60	14
	ALL-S	19	15

each interaction, players must decide whether to be altruistic or selfish. Axelrod loaded all the submitted strategies onto a computer and had each strategy play against each of the others (and also play against itself). The overall winner of this round-robin tournament was a strategy called tit-or-tat (TFT). In round one, a TFTer acts altruistically. In each subsequent round, the TFTer mimics what the other player did on the previous round. If the other player acted selfishly in round 1, the TFTer acts selfishly in round two; if the other player acts altruistically in round 1, the TFTer acts altruistically in round two.

Here I want to ignore all but two of the numerous strategies that Axelrod assembled and focus just on TFT and ALL-S, the latter of which tells a player to be selfish on each round. TFT is a conditional strategy but ALL-S is not. Let's assume the payoffs to a player on each round are the ones shown in Table 3.6. Notice that the fitter choice for Row on each round is selfishness; this is true no matter what the Column player does. In the language of game theory, selfishness "dominates" altruism. You may think that having the players interact repeatedly doesn't change anything. Isn't ALL-S automatically fitter than TFT? In fact, the shift from looking at *behaviors* one-by-one to looking at *strategies* changes the picture dramatically.

Suppose Row and Column interact for fifteen rounds. The payoff to Row depends on that player's strategy and on the strategy used by the Column player, as shown in Table 3.7. Notice that neither strategy dominates

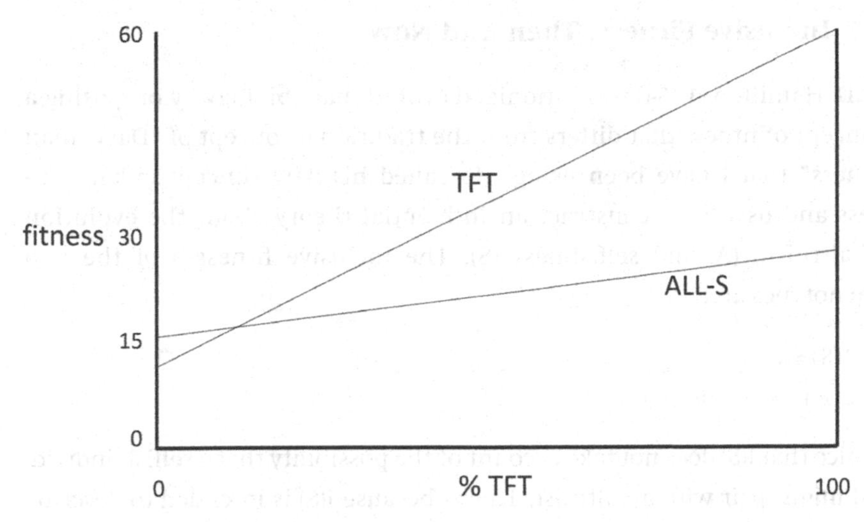

Figure 3.2 TFT versus ALL-S with 15 rounds.

the other. Now imagine a meta-population of pairs of players, where the pairs are formed at random and then play this fifteen-round game. The fitness functions for the two strategies are shown in Figure 3.2. Notice how Figure 3.2 differs from Figure 3.1. Notice also that the meta-population evolves to either 100 percent TFT or to 100 percent All-S, depending on its starting frequency. There are three types of group in this model; each group contains two ALL-S individuals, two TFTs, or one of each. In a mixed group, ALL-S outcompetes TFT. However, groups in which there are two TFTs outcompete groups in which there are two ALL-Ss. Group selection favors TFT whereas individual selection favors ALL-S. "You scratch my back, I'll scratch yours" may sound like a selfish strategy, but when TFT competes just with ALL-S, group selection is needed for TFT to evolve.

Three differences separate the one-shot game of altruism versus selfishness from the iterated game of TFT versus ALL-S. First, the additive one-shot game requires positive correlation between interactors for altruism to evolve, but TFT can evolve when pairs form at random. Second, the evolutionary endpoint in TFT versus ALL-S is either 100 percent TFT or 100 percent ALL-S in the meta-population, whereas the expected outcome in altruism versus selfishness is often a polymorphism. Finally, altruism can't evolve if groups hold together for sufficiently many generations before they send out migrants to found new groups. In contrast, the evolution of TFT is under no such pressure of time.

3.7 Inclusive Fitness, Then and Now

W.D. Hamilton (1964) revolutionized evolutionary biology by proposing a concept of fitness that differs from the traditional concept of "Darwinian fitness" that I have been using. He called his new concept *inclusive* fitness and used it to construct an influential theory about the evolution of altruism (A) and selfishness (S). The inclusive fitnesses of the two phenotypes are:

$$I(S) = x$$
$$I(A) = x - c + rb$$

Notice that I(S) does not take account of the possibility that a selfish individual might pair with an altruist. This is because I(S) is intended to describe the fitness of selfishness when altruism is very rare.[20] These inclusive fitnesses entail that

(H) $I(A) > I(S)$ precisely when $r > c / b$.

As before, c is the cost of being an altruist, b is the benefit that comes from receiving an altruist's donation, but r is something new; it is Sewall Wright's (1922) *coefficient of relationship*, which describes the genealogical relationship that two individuals bear to each other. Full siblings have an r value of 0.5 as do parent and offspring; monozygotic twins have $r = 1.0$. To see what these numbers represent, consider two diploid individuals in the same generation; they may be full sibs, or half-sibs, or first cousins, etc. If one of them has a single copy of a gene and inherited it from the most recent common ancestor of the two individuals, what is the probability that the other individual also inherited that gene from that ancestor? If they are full sibs, their most recent common ancestor is their shared parental pair, and the answer is that the probability is ½. The genes shared by the sibs are said to *identical by descent*. They aren't merely *identical by type*. The r value would be the same if the gene in question were very common.[21] The r value pertains to gene *tokens*, not to gene *types* (Section 2.5).

[20] This is Hamilton's way of addressing the problem of origination.

[21] Wright's r does not describe how similar two genetic relatives are to each other genetically. Human beings, regardless of how distantly related they are to each other, have more than 99 percent of their genes in common.

Hamilton (1964) introduced this new concept of fitness because he thought that the traditional concept of Darwinian fitness is unable to explain the evolution of helping behaviors beyond those that arise in parental care. This is a mistake, as you can see from the use of Darwinian fitness to model the evolution of altruism in Section 3.4. Hamilton viewed the inclusive fitness concept as a sane alternative to the concept of group selection, which he thought was hopelessly muddled. Hamilton soon changed his mind, however, as shown by his ground-breaking paper on sex ratio evolution (Hamilton 1967). Another reason for his change of mind about group selection was an equation derived by George Price (1970, 1972), which decomposes the total change in trait frequency caused by natural selection into two addends – the change due to individual selection and the change due to group selection.[22] The Price equation led Hamilton to view group selection as a clear mathematical concept, something worlds away from what had come before. Despite Hamilton's change of mind, many biologists still view Hamilton as the hero who slew the group selection dragon once and for all.

Hamilton (1964) notes that "r" can be reinterpreted so that it no longer measures genealogical relatedness. What is key to r, under this broader interpretation, is that it measures how often altruists tend to interact with altruists compared with how often selfish individuals tend to interact with altruists. Understood in this way, Hamilton's r corresponds to the quantity $p–q$ discussed in Section 3.5, and Hamilton's criterion H and the Darwinian criterion D are equivalent in the sense that they agree on when altruism will evolve and when it will not (Dugatkin and Reeve 1994). Without this reinterpretation of r, the H criterion sometimes produces the wrong answer to the question of whether altruism will increase in frequency.

After Hamilton's (1964) paper appeared, the standard interpretation of his model was that it describes *individual* selection, albeit one in which fitness is defined in a new way. The fact of the matter is that Hamilton's model involves groups that vary in their inclusive fitness. True, Hamilton in that paper and other inclusive fitness theorists did not use the g-word, but that doesn't show that there is no group selection in their models.

22 I'll discuss this briefly in Section 3.13.

3.8 Community Selection

The evolution of altruism by group selection is usually conceptualized as a process in which all the individuals are members of a single species, but the same logic applies to the evolution of altruism in multi-species communities (Sober and Wilson 1998, pp. 118–123). The human microbiome is arguably an example. About 39 trillion bacteria, fungi, protozoa, and viruses live on or inside each of us. These microbes come from about 10,000 different species. Some help us by stimulating the immune system, degrading toxins, and synthesizing vitamins and amino acids, but others are harmful. A baby's microbiome mostly comes from its mother. Uhr et al. (2019) report that "after around 3 years, the relative proportions of microbial taxa remain mostly stable, but the microbiome composition can be altered over time by changes in diet as well as by antibiotics." It has been argued that the multi-species community that is made of a human host and its microbiome (which together are called a *holobiont*) evolved under the influence of community selection (Dupré 2012; Gilbert, Sapp, and Tauber 2012).

Community selection experiments reveal that the process can produce evolutionary change. Swenson et al. (2000) began their remarkably simple and low-cost experiment by placing sterilized soil in a number of flowerpots and inoculating the pots with a soil sample drawn from a garden, thus introducing a multi-species community into each. The experimenters planted identical grass seeds in each pot, watered them, and then waited a number of days. They then created a new set of flowerpots with sterilized soil and inoculated them with soil drawn from the first-generation pot whose grass had grown the tallest. Then they added grass seeds. This procedure was repeated several times, with the result that the average height of the grass at the end of the experiment was much greater than the average height at the beginning.[23]

Here's a simple model of how altruism can evolve by community selection that mimics what I said about the evolution of altruism and selfishness under the influence of within-species group selection. Suppose each community has four members – two organisms from species 1 and two

[23] See Goodnight (1990) for discussion of other experiments in which there is selection on multi-species communities.

from species 2. Organisms from species 1 are either altruistic (A1) or selfish (S1) and organisms from species 2 are either altruistic (A2) or selfish (S2). Suppose further that an altruist in a given species provides a fitness benefit to the three other individuals in its quartet. As before, quartets are fitter the more altruists they contain, but an altruist does worse than a selfish individual in the same species if they are in the same quartet. Altruism can evolve in both of these species if the parameter values are right.[24]

Admirers of Hamilton's (1964) paper often regarded the coefficient of relatedness (r) as the key to understanding when altruism will evolve. That coefficient has a very low value in community selection, but altruism can evolve nonetheless. This should be enough to lay to rest the fixation on genealogical relatedness. What matters to the evolution of altruism in the additive model discussed in Section 3.4 is that altruists mostly interact with altruists and selfish individuals mostly interact with selfish individuals; close genealogical relatedness of interactors is just one way to achieve that correlation.[25] But stepping back from the additive one-shot prisoners' dilemma makes room

[24] **Question:** Darwin (1859, p. 87) says that "what natural selection cannot do, is to modify the structure of one species, without giving it any advantage, for the good of another species." Does the model of community selection just described refute Darwin's thesis?

[25] Doolittle and Inkpen (2018) agree that evolution by community selection is possible when new communities form by budding-off from old ones; this is an instance of uniparental inheritance, wherein each daughter community has just one parent community. However, they think the situation changes dramatically when

> … all parental communities release the individual organisms they contain into a common pool, from which n are randomly chosen to form each community in the next generation of communities. In this case, the species compositions of the offspring communities will reflect only that of the common pool and will, sampling error aside, resemble each other.

In this scenario each daughter community traces back to between 1 and n parent communities. Doolittle and Inkpen think there are "too many parents" when n is large enough, but how large is large enough? They correctly note that "by the law of large numbers, such intercommunity resemblance will get closer and closer as n increases." For any value of n, the variance can be substantial if the common pool has high variance. The formula for the variance among the mean values in samples of size n drawn with replacement from a population of size N, where X is the quantitative variable of interest, is var(X)/n; if there is sampling without replacement, that quotient is multiplied by (N – n)/(N – 1) (Grimmett and Stirzaker 2001, p. 87). The variance in the population goes up as more and more independent variables (each with positive variance) are considered. Doolittle and Inkpen

for the fact that correlation of interactors is not required – both when the one shot prisoner's dilemma is non-additive and when an iterated prisoners' dilemma is underway. Your relationship with your microbiome may well be non-additive and it certainly isn't one-shot; it is a continuing saga.

3.9 Group Selection without Altruism

Darwin invented the idea of group selection to deal with the problem of altruism, and altruism has dominated discussions of group selection ever since. However, there is more to group selection than the question of how altruism evolves. In *The Descent of Man*, Darwin provides a neat example:

> Now, if some one man in a tribe, more sagacious than the others, invented a new snare or weapon, or other means of attack or defence, the plainest self-interest, without the assistance of much reasoning power, would prompt the other members to imitate him; and all would thus profit … If the new invention were an important one, the tribe would increase in number, spread, and supplant other tribes (Darwin 1871, p. 161).[26]

Darwin does not give many details here, but it is easy to imagine that innovators do better than non-innovators in the same tribe. Inventors get first use of their inventions and they may also be rewarded, especially if they share or sell their inventions. The invention spreads within the tribe, not by a gene's increasing in frequency, but by individuals learning from the innovator. Later on, group selection occurs in the competition *among* tribes. In this example, the trait of interest is good for the individual *and* for the group as well.[27] Okasha (2020) says that in this type of case, "the fact that the behaviour has a beneficial effect on the fitness of others is a mere side-effect, or byproduct, and is not part of the explanation for why the behaviour evolves." I disagree.

assume in their argument that s<<n and n<p, where s is the number of species in the common pool and p is the number of parent communities that contribute to that pool. I think these assumptions aren't necessary for evolution by community selection, and also that they are insufficient to deliver the conclusion that Doolittle and Inkpen draw. The authors cite Godfrey-Smith (2012) as the source of the too-many-parents argument.

[26] My thanks to Alexander Rosenberg for drawing my attention to this passage.

[27] **Exercise:** Draw the fitness functions for Innovate and Don't Innovate. How are group and individual selection represented in your diagram? **Question:** Is getting vaccinated against an infectious disease an example of evolutionary altruism?

Table 3.8 *Payoffs to row in the hawk/dove game*

		Column	
		Dove	Hawk
Row	Dove	15	0
	Hawk	30	−10

The second example of group selection without altruism comes from evolutionary game theory. Maynard Smith and Price (1973) invented the Hawk/Dove game to show how natural selection can lead organisms to restrain themselves in their combat with conspecifics. Hawk and Dove are two strategies that the individuals in a single species might follow. Unlike their namesakes in the real world, Doves in this game are pacific while Hawks are very aggressive. When two Doves come upon a food item that is worth 30 units of fitness, they don't fight over it; they merely toss a fair coin to decide who will get the benefit, so the average Dove in this situation gets 15 units of fitness. When two Hawks face this situation, they fight aggressively; the loser gets severely injured (and the fitness result is −50), while the winner gets the food and thereby gains 30 units of fitness. The average payoff to a Hawk in this situation is (30 – 50)/2 = −10. When a Hawk and a Dove interact, the Dove retreats and gains nothing, while the Hawk gets the food without injury and gains a benefit of 30. These four payoffs are described in Table 3.8. The best thing that can happen to the Row player is to be a Hawk who is interacting with a Dove; the worst is to be a Hawk who is interacting with another Hawk.

Now imagine a population of pairs of individuals. There are three types of pair – two Hawks, two Doves, and one of each. Table 3.8 shows that Hawks do better than Doves when Hawks are rare, whereas Doves do better than Hawks when Doves are rare. The result is that a polymorphism evolves in which Hawks and Doves are both maintained in the population; natural selection drives neither trait to zero.[28]

Maynard Smith and Price thought this game shows that the hypothesis of group selection isn't needed to explain restraint in conspecific combat; they

[28] **Exercise**: Draw the fitness functions that represent Hawk and Dove and figure out what the equilibrium frequency is.

were responding to biologists like Konrad Lorenz (1963) who claimed that restraint evolved because it was good for the species. The ironic fact of the matter is that the Hawk/Dove model involves group selection, even though the inventors of the model did not use the g-word. The groups are of size 2. Individual selection occurs only in Hawk/Dove pairs, because that's the only type of group in which individuals vary in fitness. Within mixed groups, Hawk outcompete Dove. However, there is selection against Hawks at the group level, since HH groups are less fit than both DD and DH groups.[29]

3.10 A Third Unit of Selection – Intragenomic Conflict

It's now time to downshift to a unit of selection that Darwin never considered. Instead of thinking of organisms as parts of the groups in which they live, let's think of genes as parts of the organisms in which they reside. Just as individuals in the same group can be selfish or altruistic, so genes in the same organism can be selfish or altruistic. There is a sense in which this idea is old hat. If a gene causes the organism in which it lives to be selfish or altruistic in its interactions with other organisms, then it is natural to say that the gene itself is selfish or altruistic. The idea I'll now explain describes a different sense in which genes can be selfish or altruistic. We need to think of genes competing with each other *within the same organism*. This is the idea of *intragenomic conflict*.

The example of intragenomic conflict that I want to describe requires a brief review of material from high school biology. Mendel's first law of heredity (the law of segregation) says that diploid parents (so called because they have their chromosomes in pairs) produce haploid gametes (sperms and eggs that each have singleton chromosomes), which then come together in reproduction to produce diploid offspring. Gametes form by the process of meiosis, wherein pairs of chromosomes are separated (segregated) from each other. Mendel's second law (the law of independent assortment) says that AB heterozygotes produce 50 percent A-bearing gametes and 50 percent B-bearing gametes. It has been known since the 1940s that this law has exceptions. Gamete formation is often 50/50, but there are genes that distort the segregation ratio, so that a DW heterozygote produces more than 50 percent D and less than 50 percent W gametes. Here

[29] **Question:** Are Doves altruists in the Hawk/Dove game?

Table 3.9 *Effects of the t and w alleles on a target chromosome's production of viable sperm*

		State of the target's partner	
		t	w
state of the target chromosome	t	0 percent	85 percent
	w	15 percent	50 percent

"D" stands for *distortion* or *driving* and "W" stands for *wild type*.[30] Segregation distorter genes induce meiotic drive, thus breaking Mendel's second "law."

A driving gene called the t allele was discovered in the house mouse (Dunn 1953). In tw heterozygotes, 85 percent of the gametes are t and 15 percent are w. Male tt homozygotes are sterile, while ww and tw males are fertile. These details are summarized in Table 3.9, which describes the pair of chromosomes on which t and w reside. I focus on one of them, which I call "the target." Notice that the table represents a chromosome's production of gametes, not an organism's production of babies.

In heterozygotes, the t allele outcompetes the w allele in the game of forming gametes, but ww homozygotes outcompete tt homozygotes in the game of having babies. There is selection *within* heterozygote individuals and selection *between* homozygote individuals. These opposing forces both affect the evolution of the alleles. The upshot of this selection process is illustrated in Figure 3.3. When t is rare, it is almost always found in heterozygotes; t does very well then, since it achieves an 85 percent representation in the gamete pool that heterozygotes produce. However, when t is very common, it is almost always found in homozygotes, and those individuals are sterile. That's why the t line in Figure 3.3 runs downhill, from a fitness of 85 percent to a fitness of 0 percent. Now let's consider gene w. When w is common, it is mostly found in homozygotes, and fair meiosis guarantees that a copy of that gene on the target chromosome has a 50 percent representation in the gamete pool. On the other hand, when w is rare, it is found almost always in heterozygotes, and in that context w gains for itself only a 15 percent representation in the gamete pool that

[30] "Wild type" is a term from genetics; it refers to an allele that is prevalent in the population before some gene of interest arises by mutation.

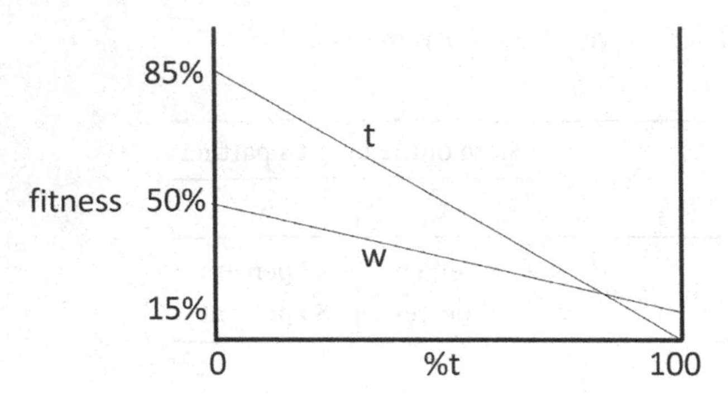

Figure 3.3 Meiotic drive in the house mouse.

heterozygotes produce. That's why the w line in Figure 3.3 also runs downhill. Notice that when t is rare, t is fitter than w, and when w is rare, w is fitter than t. Using the rule of thumb that fitter traits increase in frequency and less fit traits decline when selection is in control, you can see that the population will evolve to an intermediate frequency, which is located at the point in Figure 3.3 where the two fitness lines cross.[31]

A nice feature of this simple model of the t-allele is that it makes a precise prediction of what the frequency of t will be when a house-mouse population reaches its equilibrium. The prediction is that t should be present at 85 percent. Lewontin and Dunn (1963) observed that the frequency is lower, and suggested that a third force must have affected the gene's evolution. In addition to intragenomic conflict and individual selection, they conjectured that group selection played a role. Groups go extinct when all the males in the group are tt homozygotes, and females in those groups will likely contain high frequencies of the t allele. Intragenomic selection promotes the evolution of the t allele, but individual and group selection each push in the opposite direction.[32,33]

[31] CRISPR gene drives is a new technology for manipulating genes (Doudna and Charpentier 2014); it mimics meiotic drive alleles found in nature. CRISPR gene drives are now being developed with the goal of eradicating malaria by changing genes in the mosquito species that spread the disease from one human being to another (Bier and Sober 2020).

[32] I mentioned earlier that group selection can influence the evolution of biased sex ratios. Meiotic drive does the same thing when driving genes are on sex chromosomes; for an interesting example involving *Drosophila*, see Muirhead and Presgraves (2021).

[33] **Question:** G. C. Williams's 1966 book *Adaptation and Natural Selection* was a powerful and influential critique of group selection hypotheses. He cites Lewontin and Dunn

Driving genes like the t allele deserve to be called "selfish genes" since they promote their own spread and impose costs on the organisms in which they live. They resemble successful parasites. This may sound like Richard Dawkins' (1976) idea of a selfish gene, but Dawkins used the term "selfish" in a much more expansive way. For him, any gene that evolves by natural selection is selfish. This means that when a gene evolves because it causes organisms to sacrifice their own fitness and augment the fitnesses of others, Dawkins will say that the gene is selfish. This turns selfishness into a useless category. The "survival of the fittest" is not a tautology when fitness is understood as an ability (Section 2.2), but "only selfish traits evolve by natural selection" is a tautology, given Dawkins's usage.

There's an important lesson to be learned from intragenomic conflict. The harmonious cooperation of genes with each other to promote the survival and reproduction of the organisms in which they reside is something that needs to be explained. It is not an automatic given. Just as organisms live in groups, so too do genes live in organisms. Lower-level entities can subvert the interests of higher-level entities. It takes higher-level selection to diminish the impact of *subversion from within*.[34]

3.11 Species Selection and Cultural Group Selection

There are two further types of selection that require changes in the conceptual details that figure in the treatment of units of selection described thus far. The first is *species selection*. The idea here is that species sometimes differ in their abilities to give birth to daughter species (Eldredge and Gould 1972, Stanley 1975 and 1979, Vrba 1980, Maynard Smith 1989). What is conceptually interesting in this hypothetical process is that the reproductive success of a species is computed by counting the number of daughter *species* it produces. This contrasts with models of group selection in which group productivity is measured by seeing how many daughter *organisms* the group produces. The "head-counting paradigm" that is familiar from standard conceptions of natural selection (wherein you measure reproductive success by counting babies, not the number of cells in

(1963) as the only convincing published argument for the influence of group selection. Should Williams have concluded from the Lewontin and Dunn study that there probably are lots of traits whose evolution was influenced by group selection?

[34] This nice phrase is from Dawkins (1976).

babies, nor do you measure the total biomass of offspring) thus needs to be supplemented (Sober 1984).[35]

The second novelty we need to recognize takes account of the fact that teaching and learning can cause individuals in the next generation to resemble those in the one before. Human beings acquire beliefs and preferences from members of the previous generation, but those ideas aren't caused by the genes that individuals receive from their biological parents. For example, there is no gene for speaking one language rather than another, and yet children often speak the same language as their biological parents. Cultural transmission is different from genetic transmission. As Darwin noted in his discussion of the evolution of morality, and also in his discussion of technological innovation, groups of individuals can differ culturally in ways that affect the abilities of groups to compete with each other. The result is *cultural group selection* (Cavalli-Sforza and Feldman 1981; Boyd and Richerson 1985; Lewens 2015).[36] The conflict between Nuer and Dinka in South Sudan provides an example (Sober and Wilson 1998). This conflict has been under way since at least the nineteenth century. Each of these tribes contains numerous local groups. Nuer individuals in the same group cooperate with each other, and Nuer groups help other Nuer groups when they are attacked by Dinkas. The Dinka had less cooperation within and between local groups. Each tribe kidnapped children from the other and raised them as full-fledged group members, so there was little or no genetic distinctness between the two. When Sober and Wilson described this example, the Nuer were winning, owing to the cultural differences between the two groups, but the story did not end then. The bloody, sometimes genocidal, conflict continues.

As this sad example illustrates, group selection theory does not see the world through rose-colored glasses; group selection is not a process that automatically produces universal niceness. A better slogan is that group selection produces within-group niceness and between-group nastiness (Sober and Wilson 1998).[37] Just as individual selection can leave

[35] Damuth and Heisler (1988) use "MLS1" and "MLS2" to label these two ways of defining group fitness in the context of MLS theory.

[36] Notice that cultural group selection can fail to result in evolution if "evolution" is defined as change in the genetic composition of populations (Section 1.3). I'll discuss an example of cultural evolution that doesn't involve selection in Section 4.1.

[37] This slogan is better, but it isn't perfect. To see why, consider the fact that it is logically possible for a human group in which there is slavery to outcompete another group in which there is not.

individuals worse off at the end of the process than they were at the beginning (Section 2.8), group selection can leave groups worse off at the end of the process than they were at the beginning. How many wars in human history have been like this? Group selection is a competitive process just as much as individual selection is. The difference is that the locus of competition has been shifted to a different level of organization.

3.12 Reductionism

Reductionism is a long-standing topic in philosophy of science.[38] It often involves an asymmetry thesis about *macro* and *micro*. For example, in philosophy of social science, *methodological individualism* is the reductionist thesis that when societies change, the best (or only) explanation describes what individual people think and do; appeal to social facts to explain societal changes is at best a stopgap expedient (Wright et al. 1993; Heath 2020). Here societies are the macro objects and individual people are the micro.[39]

To see how questions about reductionism arise in the units of selection problem, I'll formulate an example by using the algebra from Section 3.4.[40] Suppose that altruism (A) increases in frequency and selfishness (S) declines in a meta-population in which groups are of size 2. Here are three possible explanations of why that happened:

(M) There was group selection for A, individual selection for S, and the former was more powerful than the latter.

(I-1) Individuals with trait A were on average fitter than individuals with trait S.

(I-2) Individuals with trait A almost always lived with other individuals who had trait A (so A individuals almost always had a fitness of $x - c + b$ and only rarely had a fitness of $x - c$), and individuals with trait S almost always lived with other individuals with trait S (so S individuals almost always had a fitness of x and only rarely had a fitness of $x + b$), and $b >> c$.

[38] For an overview of reductionism in biology, see Brigandt and Love (2017).

[39] Micro-reductionism leads by a transitivity argument to the conclusion that the best explanation of an event is in terms of particle physics. **Question:** What does "transitivity argument" mean here?

[40] For more general discussion, see Wimsatt (1980), Sober (1980b), and Okasha (2006).

Reductionists argue that the MLS explanation M is inferior to one or both of the individualistic explanations I-1 and I-2. Are they right?

Looking first at M and I-1, I suggest that M is the better explanation if you want more causally relevant details, but I-1 is better if you want fewer (Sober 1999).[41] However, in neither case is what you want a reason to declare M false and I-1 true. Friends of I-1 should recognize that the truth of I-1 does not entail that M is false, and friends of M should recognize that the truth of M does not justify rejecting I-1. In fact, I-1 must be true if M is. I-1 describes the net causal impact of two component causes. Here's an analogy: If Sam pushes a billiard ball east at the same time that Aaron pushes it west and Aaron's push is harder, the resultant of these two vectors is a push to the west. When the ball moves west, you can explain why by citing the one net cause or the two components.

Looking next at M and I-2, I suggest that both are good explanations, and they are not in conflict. In fact, given how groups should be defined in discussions of group selection (Section 3.5), I-2 entails M, which means that I-2 is not a reason to reject M. True, I-2 does not use the g-word, but I-2 entails the existence of group selection nonetheless.

A second question about reductionism can be posed about units of selection. Instead of asking which of two explanations is better, you can ask whether one explanation reduces to another. Group selection does not reduce to individual selection, since you can't deduce that group selection is occurring just by describing facts about individual selection. This is because the variance in fitness within groups leaves open whether there is variance in fitness among groups. Using vocabulary from Section 2.4, you can say that group selection does not *supervene* on individual selection. However, there is a second reductionist claim that is correct: group fitnesses supervene on individual fitnesses if the productivity of a group is the sum of the productivities of the organisms in the group. Here it is worth remembering the discussion in Section 3.10 of the head-counting paradigm and the possibility of defining group fitness as the expected number of daughter *groups* that a group founds. If group fitness is defined in that second way, group fitness does not reduce to individual fitness (Okasha 2006).

[41] Friends of M might be inclined to say that I-1 is completely trivial, but this is a mistake, since I-1 rules out the possibility that A's evolution was due to drift and to drift alone. Drift is the subject of the next chapter.

3.13 Two Conventionalisms[42]

You can see what a unit of selection is by generalizing from the three main examples of units of selection discussed in this chapter. When intragenomic conflict affects the evolution of trait T, the gene is a unit of selection in the evolution of trait T. When selection among the individuals within a group affects the evolution of T, the individual is a unit of selection in the evolution of T. And when selection among the groups within a meta-population affects the evolution of trait T, the group is a unit of selection in the evolution of T. In each case, selection at a level exists precisely when there is variation in fitness at that level.[43] These definitions allow for the possibility that different traits evolve because of different causes, and it also allows for the possibility that a given trait evolves because of multiple causes. This double pluralism stands in contrast with *genic selectionism*, a term often used for the monistic thesis that the gene is the one and only unit of selection, no matter what trait you consider (Lloyd 2020).

MLS theory embodies a *realist* view of the units of selection problem. By "realist" I mean that MLS theory takes it to be a factual question about nature what the units of selection are that affect the evolution of a given trait. The units of selection that are involved in the evolution of trait T are as real as the patterns of fitness variation with respect to trait T that exist at multiple levels of organization. The opposite of realism is *conventionalism*, and it has taken two forms. First, several philosophers have argued that it is a matter of convention whether you explain the evolution of a trait by invoking group selection, individual selection, or genic selection. The second kind of conventionalism emerges as a possible interpretation of the work of biologists who developed quantitative frameworks for representing how much of the total change in trait frequency caused by natural selection on a trait is caused by group selection and how much is caused by individual selection. These frameworks agree that group and individual selection are conceptually distinct, but they define those two causal processes differently. Biologists who argue about

[42] Much of the prose in this section is from Sober (2011c).

[43] **Exercise:** Write a general definition of "the X is a unit of selection in the evolution of trait T."

which of these frameworks is better are often realists about the question they are addressing, but the possibility must be considered that the choice between the two is a matter of convention.

The conventionalist thesis endorsed by philosophers Kim Sterelny and Philip Kitcher (1988), Kitcher, Sterelny, and Kenneth Waters (1990), Waters (1991, 2005), and Sterelny and Paul Griffiths (1999) asserts that any trait that evolves by natural selection can be said to evolve by genic selection, whereas only some of the traits that evolve by natural selection evolve by group selection or individual selection. They conclude from this that explanations that appeal to individual and group selection are *optional* and that genic selection explanations are *better* because the genic selection framework is more general.[44] This greater generality is said to follow from the thesis that evolution is change in gene frequencies, which means that evolution by natural selection must involve some genes being fitter than others. This conventionalism differs from the position promoted by early critics of group selection hypotheses (Williams 1966, Maynard Smith 1976, and Dawkins 1975). These critics were realists, holding that group selection hypotheses are factually mistaken claims about nature.[45]

I endorsed explanatory pluralism in the previous section by discussing altruism and selfishness, but that was just an example. My more general thesis is this: if an event has one true causal explanation, there must be many. That multiplicity involves both time and space. Some explanations invoke proximate causes, while others invoke distal, and some explanations mention macro causes while others mention micro. An additional way to see this multiplicity of explanations is to recognize that good explanations can be more general or more specific. The former abstract away from explanatory details whereas the latter do the opposite. This pluralism

[44] Notice that the appeal to generality that conventionalists make here is different from the reductionist's claim that micro-explanations are better than macro.

[45] Sterelny and Kitcher (1988) liken their conventionalism to a form of conventionalism discussed in philosophy of physics – that the choice between Euclidean and non-Euclidean geometry as theories of the structure of physical space is a matter of convention. I think there is a big difference. The claim that space is Euclidean and the claim that space is non-Euclidean are *incompatible* with each other. Conventionalism asserts that there is *no fact of the matter* as to which is true. This is different from the epistemic claim that it's impossible to tell which geometry is true. Kitcher and Sterelny think that "competing" hypotheses about units of selection are compatible with each other; in fact, they think that they are connected by entailment relations, as I'll explain.

about explanation may seem to entail conventionalism about units of selection, but it does not. To see why, let's walk through the philosophers' conventionalist argument more carefully.

If the groups in a meta-population vary in fitness because they have different frequencies of trait T and the frequency of T in the meta-population change for that reason, then it must also be true that individuals that have T and individuals that have not-T have different average fitnesses in the meta-population. Conventionalists then define individual selection as variation in fitness among individuals in the meta-population and declare a victory, pointing out that you can say that the traits evolved by group selection if you want to, but you can also say that the traits evolved by individual selection. Both explanations are true, so the choice between them is said to be a matter of convention.

These conventionalists deploy the same line of argument one level down when they compare individual and genic selection hypotheses. If individual selection causes trait frequencies in a meta-population to change, and this change counts as genuine evolution (defined as change in gene frequencies), then genes in the meta-population must have differed in fitness. Conventionalists then define genic selection to mean variation in the fitnesses of genes in the meta-population. They conclude that a trait that evolves by individual selection can also be said to evolve by genic selection. Once again, the choice is said to be a matter of convention.[46]

The point to notice about this conventionalist argument is that MLS theory and conventionalism use different definitions of individual and genic selection, as shown in Table 3.10. MLS theory takes group, individual, and genic selection to be logically independent of each other, while the conventionalism now under discussion defines the three so that they are linked by entailment relations (with higher-level selection entailing lower-level selection, but not conversely). This polysemy is exasperating, but does it hold out the hope that we can all be friends? Well, I am happy to be a conventionalist about the descriptors that conventionalists use. However, I don't see why conventionalists are entitled to take a conventionalist view of the distinctions that MLS realists draw.

[46] The conventionalism described here is not reductionistic. It says that "a trait evolved by group or individual selection" logically entails that the trait evolved by genic selection, but not conversely. If H entails L but not conversely, it is false that H reduces to L.

Table 3.10 *MLS theory and conventionalism define two of these three levels of selection differently*

	MLS Theory	Conventionalism
Group selection	Fitness variation among the groups in the meta-population	Fitness variation among the groups in the meta-population
Individual selection	Fitness variation among the organisms in a group	Fitness variation among the organisms in the meta-population
Genic selection	Fitness variation among the genes in an organism	Fitness variation among the genes in the meta-population

I turn now to a different debate, this one initiated by biologists. Throughout this chapter, I've described what it means for group and individual selection to occur and what it means for one of them to be stronger than the other when they push in opposite directions. I never proposed a quantitative criterion for how the total change in a trait's frequency induced by natural selection should be partitioned into the amount of change induced by each of the two causes. George Price (1972) developed a quantitative solution to this problem in what is now called the Price equation. An alternative framework called "contextual analysis" was worked out by I. Loraine Heisler and John Damuth (1987); they showed how a methodology used in the social sciences can be applied to the units of selection problem. As a result, there now are two versions of MLS theory that offer different ways of quantifying the intensity of group and individual selection.

Although the Price equation and contextual analysis make identical predictions concerning which traits will increase in frequency, they assign different meanings to group and individual selection. Two examples of this divergence are depicted in Figure 3.4. There is "soft selection" in the example on the left, meaning that the average fitnesses of groups (as represented by their values for $\bar{w}$) do not vary, though in each of them selfish individuals are fitter than altruists. The Price approach says there is no group selection here, as there is no variation in fitness among groups. Contextual analysis says there is, since an individual's fitness depends on what sort of group the individual is in. In the right-hand part of Figure 3.4, the fitnesses of Fast and Slow (running speeds in zebras) have a very slight degree of frequency dependence; their fitness lines have a tiny negative

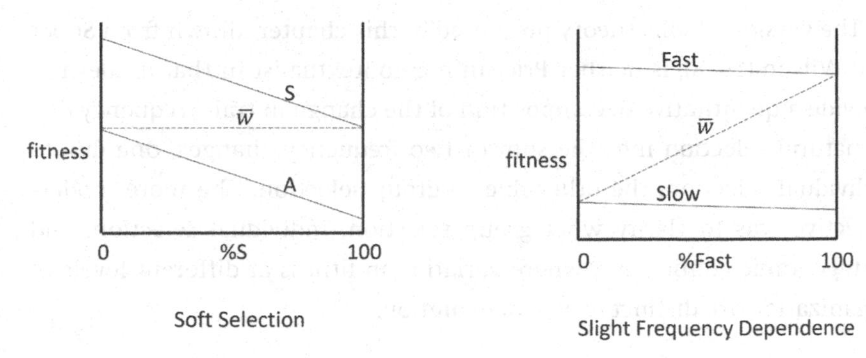

Figure 3.4 The Price approach and contextual analysis classify these two examples differently.

slope. Contextual analysis says that there is very little group selection here, since an individual's fitness depends hardly at all on the kind of group it is in. The Price approach says that the intensity of group selection depends on how much the groups in the meta-population differ in fitness. That isn't depicted in the figure, but suppose that fast zebras are very rare in half the groups in the meta-population and are very common in the other half. In that case the Price approach concludes that there is strong group selection.

You can see from these examples how contextual analysis defines group and individual selection. It says that group selection occurs precisely when an organism's fitness depends on the traits of other individuals in the same group, and that individual selection occurs when an organism's fitness depends on its own phenotype. Notice that these two definitions say nothing about how *groups* differ in fitness. Groups are simply treated as contexts that affect the fitnesses of individuals; groups are like the weather affecting the evolution of bear fur. To me this isn't group selection; it isn't selection *of groups*. Indeed, contextualism entails that group selection can occur when there is just one group.

My criticism of contextualism is not that it fails to elucidate causation. Nor do I mean that the Price approach is beyond reproach; the framework does make use of several idealizations. Okasha (2006) and Okasha and Otsuka (2020) criticize the Price equation for mistaking fortuitous group benefits for group adaptations; I think that problem is solved when the concept of a group is understood by using Wilson's (1975) idea of trait groups (Section 3.5). The trait-group concept introduces a causal consideration that is not part of the Price equation itself.[47]

[47] For further discussion of conventionalism iabout units of selection, see Bourrat (2021).

The version of MLS theory presented in this chapter, drawn from Sober and Wilson (1998), is neither Pricean nor contextualist in that it does not provide a quantitative decomposition of the change in trait frequency due to natural selection into the sum of two frequency changes, one due to individual selection, the other due to group selection. The more modest objective was to clarify what group selection, individual selection, and intragenomic conflict are, where variation in fitness at different levels of organization are distinct causes of evolution.

4 Common Ancestry

4.1 Darwin on Common Ancestry

In the closing pages of the *Origin*, Darwin offers a summary of his views about common ancestry:

> I believe that animals have descended from at most only four or five progenitors, and plants from an equal or lesser number. Analogy would lead me one step further, namely to the belief that all animals and plants have descended from some one prototype. But analogy may be a deceitful guide. Nevertheless all living things have much in common, in their chemical composition, their germinal vesicles, their cellular structure, and their laws of growth and reproduction. We see this even in so trifling a circumstance as that the same poison often similarly affects plants and animals; or that the poison secreted by the gall-fly produces monstrous growths on the wild rose or oak-tree. Therefore I should infer from analogy that probably all organic beings which have ever lived on this earth have descended from some one primordial form, into which life was first breathed.[1] (Darwin 1859, p. 484)

Darwin's idea that the hypothesis of universal common ancestry is justified by the fact that "all living things have much in common" can be seen to derive from two others – that when two or more taxa have trait T, this "matching" favors the hypothesis of common ancestry over the hypothesis of separate ancestry,[2] and the more the matches outnumber the mismatches, the stronger the favoring.

[1] **Question:** Given that the thesis of universal common ancestry is not about the number of start-ups (Section 1.1), did Darwin err when he wrote that all organisms today are descended from a "primordial form, into which life was first breathed"?

[2] The "matching" I'm talking about here means merely that the two taxa both have trait T; they need not share *all* their traits. The traits that biologists cite as evidence for

Darwin advances a second epistemological thesis about common ancestry – that some matches provide stronger evidence for common ancestry than others:

> ... adaptive characters, although of the utmost importance to the welfare of the being, are almost valueless to the systematist. For animals belonging to two most distinct lines of descent, may readily become adapted to similar conditions, and thus assume a close external resemblance; but such resemblances will not reveal – will rather tend to conceal their blood-relationship to their proper lines of descent (Darwin 1859, p. 427).

On the next page, he gives the example (noted in Section 1.9) of the "shape of the body and the fin-like limbs" found in whales and fishes; these are "adaptations in both classes for swimming through water" and thus provide little or no evidence that the two groups have a common ancestor.

Darwin's claim that adaptive similarities provide scant evidence for common ancestry sounds right when it is applied to his example of whales and fishes, but there are other examples that make it sound wrong. Here's how Darwin describes one of them:

> The framework of bones being the same in the hand of a man, wing of a bat, fin of the porpoise, and leg of the horse – the same number of vertebrae forming the neck of the giraffe and of the elephant, – and innumerable other such facts, at once explain themselves on the theory of descent with slow and slight successive modifications. The similarity of pattern in the wing and leg of a bat, though used for such different purposes, – in the jaws and legs of a crab, – in the petals, stamens, and pistils of a flower, is likewise intelligible on the view of the gradual modification of parts or organs, which were alike in the early progenitor of each class (Darwin 1859, p. 479).

Darwin believes that the shared "framework of bones" is strong evidence for common ancestry, and yet he also thinks that this morphology is useful in the different groups, so which epistemological principle is right – that adaptive similarities *always* provide weak evidence for common ancestry, or that *some* adaptive similarities provide weak evidence

common ancestry are usually traits of individual organisms; when they say that a taxon has trait T, they usually mean that all or almost all the individuals in the taxon have that trait.

while others provide strong? If the latter, what distinguishes the weak from the strong?

Darwin's wording suggests an answer to this last question: perhaps the shared framework of bones is strong evidence for common ancestry because it is used for different purposes in these different groups. This suggestion seems to separate the torpedo-shape of whales and fishes from the limb morphology of human beings, bats, porpoises, and horses. However, a more modern example casts doubt on this diagnosis. Francis Crick (1957) took what was then thought to be the universality of the genetic code[3] to be strong evidence for universal common ancestry. Modern biology has retained his conclusion even though it now is known that the genetic code isn't universal; it is *nearly* universal, with almost all groups of organisms using one code and a few others using codes that are slightly different (Knight et al. 2001). The prevalent genetic code provides strong evidence for common ancestry even though it has the same function in all the living things that have it. The puzzle remains – how to separate adaptive similarities that provide strong evidence for common ancestry from adaptive similarities that do not?

4.2 What Is Ancestry?

Before exploring the epistemological questions just described, I want to clarify what common ancestry means. Starting with the idea of ancestry itself, we need to understand the nature of the relationship that connects ancestor to descendant. The relationship exists in virtue of a chain of parent/offspring relationships. That is, A_1 is an ancestor of A_n precisely when A_1 is a parent of A_n or there exist individuals A_2, A_3, ..., and A_{n-1} such that A_1 is a parent of A_2, A_2 is a parent of A_3, ..., and A_{n-1} is a parent of A_n.

This raises the question of what the parent/offspring relationship is. Parents cause their offspring to come into existence, but ancestry goes beyond that causal fact. Striking a match causes a flame to come into existence, but the striking isn't (literally) an ancestor of the flame. Parents cause their offspring to come into existence by a special kind of causal process – *reproduction*.

[3] The genetic code is the set of rules that cells obey that specifies how different triplets of DNA nucleotides are used to construct amino acids, which are the building blocks of proteins.

Organisms aren't the only thing that reproduce. Genes *replicate*, or perhaps the passive voice is better: genes are *replicated*.[4] A gene in an offspring typically traces back to the same (type of) gene in its parents. I say "typically" because I'm ignoring mutation and horizontal gene transfer. In the former case, a mutant gene does not trace back to genes of the same type in the parents; the gene is brought into existence by mutagens such as radiation. In the latter, genes in an organism trace back to genes that are not in the organism's parents. For example, a bacterium might insert a gene into the cell of a plant or animal. A second example of non-organisms reproducing involves groups. Groups can be related to each other genealogically. If new groups form by budding off from old ones, the reproduction is uniparental. However, a daughter group can be founded by individuals that derive from different parental groups, in which case group reproduction is multiparental.[5]

Even if you restrict your attention to the genes in an organism that come from its parents, the fact remains that different genes in the same organism can have different genealogies. The genes in human mitochondria are maternally inherited; the genes in Y chromosomes are paternally inherited. More generally, each of your genes traces back to one of your two parents, to one of your four grandparents, and so on. You have a single organismic genealogy, but the genes in your body have different genealogies.

When two organisms have a common ancestor, there exists a token organism that was the ancestor of both. For asexual organisms that reproduce by binary fission, two organisms will have a single *chain* of common ancestors, but only one of those common ancestors is their most recent common ancestor. Sexual organisms are different. An organism and its full sibling share the same mother and father, so they have *two* most recent common ancestors.

What do evolutionary biologists mean when they say that *all current organisms on earth* have a common ancestor? Biologists often take this to mean that current living things trace back to a group of primitive cells that interacted with each other, exchanging genetic material (Woese 1998, 2000). This is the second of the three possibilities depicted in Figure 4.1 (Velasco 2018).

[4] I am being fussy here because genes make copies of themselves no more than a piece of paper makes a copy of itself when placed in a copying machine. In both cases, there is machinery that produces copies.

[5] These examples don't provide a general definition of what reproduction is. I won't try to provide one.

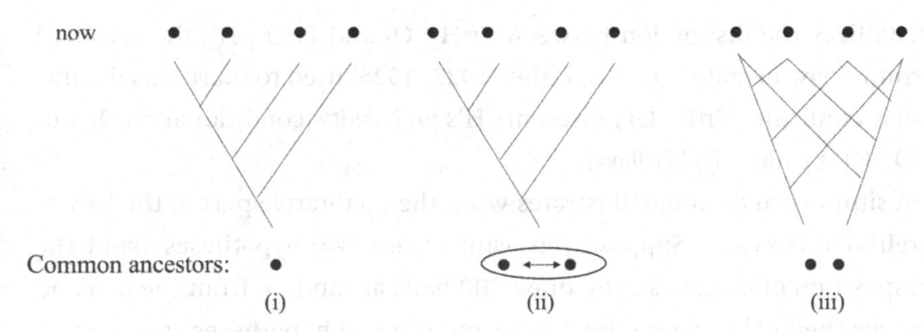

Figure 4.1 Three ways the hypothesis of universal common ancestry could be true.

In (i) and (ii), there is a single *object* (e.g., a cell or a group of interacting cells) that is a universal common ancestor; in (iii) there are two. What they have in common is the idea that all living things on the planet trace back to at least one object that is their common ancestor. Whether that object is a "species" or an "organism" depends on how those terms are defined.[6] What is ruled out is a unique universal common ancestor's being a collection of objects that never interacted, with some being ancestors of some current organisms and others being ancestors of the rest; this is *separate* ancestry.

4.3 The Law of Likelihood

To analyze Darwin's ideas concerning evidence for common ancestry, I'll use a two-part epistemological principle:

(QUAL) the qualitative Law of Likelihood: Observation O favors hypothesis H_1 over hypothesis H_2 if and only if $Pr(O \mid H_1) > Pr(O \mid H_2)$.

(QUANT) the quantitative Law of Likelihood: the degree to which O favors H_1 over H_2 is given by the likelihood ratio $\frac{Pr(O \mid H_1)}{Pr(O \mid H_2)}$.

The conditional probabilities in these two principles describe the probabilities that an observation has given this or that hypothesis, *not* the probabilities that hypotheses have given an observation. This important distinction can be seen clearly by considering an example. Suppose you hear noises coming from your attic. That's your observation O. You then consider hypothesis H, that there are noisy gremlins bowling in your attic. $Pr(O \mid H)$ is high, but I hope you agree that $Pr(H \mid O)$ is not. Although this example

[6] I discuss species in Chapter 7.

crystallizes the distinction between Pr(H | O) and Pr(O | H), the standard terminology (invented by R. A. Fisher 1922, 1925) used to mark this distinction is confusing. Pr(H | O) represents H's *probability* (conditional on O), but Pr(O | H) is called H's *likelihood*.

A simple urn example illustrates what the qualitative part of the Law of Likelihood is saying. Suppose you want to test two hypotheses about the composition of an urn, so you draw 100 balls at random from the urn and observe their color. Your observation and the two hypotheses are:

(O_1) 91 balls out of 100 are green.
(H_1) 80 percent of the balls in the urn are green.
(H_2) 40 percent of the balls in the urn are green.

I hope it makes sense to you that O_1 favors H_1 over H_2. The reason is that if H_1 were true, O_1 would be more probable than it would be if H_2 were true. Your observations don't *guarantee* that H_1 is true and that H_2 is false. In fact, they don't even guarantee that H_1 is more probable than H_2. Why not? The answer is to be found in the following equation:

$$\text{(ODDS)} \quad \frac{\Pr(H_1|O)}{\Pr(H_2|O)} = \frac{\Pr(O|H_1)}{\Pr(O|H_2)} \frac{\Pr(H_1)}{\Pr(H_2)}$$

This is *the odds formulation of Bayes' theorem*, which follows from Bayes' theorem. ODDS says that the ratio of posterior probabilities equals the likelihood ratio times the ratio of prior probabilities. I hope the temporal words "posterior" and "prior" are suggestive. The rough idea is that the priors are the probabilities the hypotheses have *before* you make your observation; the posteriors are the probabilities they have *after*. In the urn example, you know that the likelihood ratio is greater than 1, but I said nothing about the prior probabilities. That means you can't tell from the example whether the posterior ratio is greater than one.

Hacking (1965) and Edwards (1972) formulated the Law of Likelihood by using the two-place relation of support, saying that O supports H_1 more than O supports H_2 precisely when $\Pr(O \mid H_1) > \Pr(O \mid H_2)$. I use the term "favoring" because I think the relevant epistemic relation is three-place. In my view, the Law of Likelihood is irreducibly *contrastive*.

Thus far I've discussed the qualitative part of the Law of Likelihood. Now let's turn to the quantitative. Suppose you draw a second sample of 100 balls from the urn and the result is:

(O_2) 75 of the100 balls drawn are green.

Like your first observation O_1, O_2 also favors H_1 over H_2. However, there's a quantitative difference in how much the two observations favor H_1 over H_2. It is here that the likelihood ratio is useful, since

$$\frac{\Pr(O_1|H_1)}{\Pr(O_1|H_2)} > \frac{\Pr(O_2|H_1)}{\Pr(O_2|H_2)} > 1.0$$

O_1 favors H_1 over H_2 more strongly than O_2 favors H_1 over H_2.

Here's how I'll use the two parts of the Law of Likelihood in this chapter. In Section 4.3, I'll describe plausible assumptions that entail this inequality:

$$\Pr(\text{A and B have trait T} \mid \text{CA}) > \Pr(\text{A and B have trait T} \mid \text{SA})$$

Here CA is the hypothesis that A and B have a common ancestor and SA (the hypothesis of "separate ancestry") is the hypothesis that they do not. A and B can be two token organisms, or two populations. The proof of the above inequality doesn't require QUAL, but the question then arises of what this inequality tells you about the "evidential significance" of the observation that A and B have trait T. It is here that QUAL plays a role.

The assumptions that suffice to establish the above likelihood inequality leave it open that the type of evolutionary process occurring in the lineages leading to A and B may affect the degree to which A and B's having trait T favors the one hypothesis over the other. I'll address that topic in Section 4.5, using QUANT. This will involve comparing two likelihood ratios:

$$\frac{\Pr_M(A \text{ and } B \text{ have trait } T \mid CA)}{\Pr_M(A \text{ and } B \text{ have trait } T \mid SA)} > \frac{\Pr_N(A \text{ and } B \text{ have trait } T \mid CA)}{\Pr_N(A \text{ and } B \text{ have trait } T \mid SA)}.$$

Here M and N are models of different evolutionary processes. Subscripts indicate that the probability function takes a model of M or of N as an assumption. For example, if M represents the idea that there was selection *against* trait T in the lineages leading to A and B, and N means that there was selection *for* trait T in those lineages, we have a formal representation of Darwin's idea that deleterious similarities provide stronger evidence for common ancestry than adaptive similarities provide.

Before applying the Law of Likelihood to the common and separate ancestry hypotheses, I want to briefly discuss its justification. You need to recognize at the outset that the qualitative part of the Law of Likelihood doesn't perfectly capture what "favoring" means in English. When climate scientists use observations from the present and past to predict that average temperatures will probably increase in the next decade, you might summarize their assessment by saying that their observations favor the hypothesis that temperatures will rise over the hypothesis that temperatures will not. However, these scientists are asserting the following inequality:

Pr(average temperatures in next decade will be higher | present data) > Pr(average temperatures in next decade will not be higher | present data).

This inequality has the form Pr(H | O) > Pr(notH | O); it concerns the posterior probabilities of hypotheses, and those two probabilities sum to one. In contrast, the likelihoods in the inequality "$Pr(O \mid H_1) > Pr(O \mid H_2)$" need not sum to one.

Unfortunately, there is no word in English that perfectly expresses the epistemological significance of likelihood comparisons. This does not matter, however, since the point of the Law of Likelihood is not to capture what this or that word means in English. The point is to describe how likelihood inequalities and likelihood ratios are epistemically relevant. I'll use "favoring" to label that epistemic significance, with the understanding that the word sometimes has other meanings.

As mentioned, the Law of Likelihood was first formulated by R. A. Fisher (1922, 1925). He, like more recent defenders of the law (Hacking 1965; Edwards 1972; Royall 1997; Sober 2008, 2015) aren't Bayesians, since they think there is often no way to justify the assignment of prior probabilities to theories.[7] Even so, it's important to see that the Law has a Bayesian rationale. It comes from the odds formulation of Bayes' theorem. ODDS says that there is exactly one way that observation O can lead you to change how confident you are in H_1 as compared with H_2. If the ratio of posteriors is to differ from the ratio of priors, the likelihoods must differ. The more the likelihood ratio departs from 1 (either by going up or by going down), the more the ratio of posterior probabilities deviates from the ratio of priors. This is why Bayesians should embrace the Law of Likelihood.

[7] There is another reason for not being a Bayesian – there is often no way to justify the assignment of likelihoods to "catchall hypotheses." For example, consider the fact that Newton's theory of gravitation and Einstein's theory of general relativity each assigns a probability to Eddington's data on the solar eclipse he observed. However, the negation of each of these specific theories does not. Each of those negations covers *all* alternatives (it "catches them all"), including ones that have not been formulated.

Statisticians and scientists who call themselves "Bayesians" routinely do so, as revealed by the fact that they call the likelihood ratio "the Bayes factor" and think it is the crucial consideration in assessing how evidence discriminates among competing hypotheses.[8]

The Bayesian argument just given for QUAL is "top-down," in that it aims to justify the Law of Likelihood by deriving it from a more general and fundamental consideration (Bayes' theorem). This strategy does not preclude "bottom-up" arguments for QUAL that proceed by considering examples in which a body of evidence seems to favor one hypothesis over another, and then showing that the Law of Likelihood accords with those intuitive judgments. I provided a bottom-up argument in my previous discussion of the urn example. Royall (1997) provides other examples of this kind, beginning with this extreme case: suppose H_1 entails that O is true while H_2 entails that O is false. In this example, observing whether O

[8] Fitelson (2011, p. 669) says that Bayesians should reject the Law of Likelihood because favoring must have the following meaning for Bayesians:

(F) O favors H_1 over H_2 precisely when $DOC(O,H_1) > DOC(O,H_2)$

Here DOC(O,H) is a quantitative measure of the degree to which O confirms H. A Bayesian measure of degree of confirmation must conform to the following Bayesian definitions:

O confirms H precisely when $Pr(H \mid O) > Pr(H)$
O disconfirms H precisely when $Pr(H \mid O) < Pr(H)$
O is confirmationally irrelevant to H precisely when $Pr(H \mid O) = Pr(H)$

Unfortunately, there are many measures that satisfy these three conditions. Among them are:

the difference measure: $DOC_D(O,H) = Pr(H \mid O) - Pr(H)$

the ratio measure: $DOC_R(O,H) = \dfrac{Pr(H \mid O)}{Pr(H)}$

the likelihood ratio measure: $DOC_{LR}(O,H) = \dfrac{Pr(O \mid H)}{Pr(O \mid notH)}$

These and other measures differ substantively since they are not ordinally equivalent. Fitelson notes that the Law of Likelihood follows from the F condition if you use the ratio measure. However, Fitelson and other Bayesians reject the ratio measure because it entails *commutativity*, meaning that it satisfies this condition:

For any propositions A and B, DOC(A,B) = DOC(B,A).

Fitelson says that commutativity is "clearly incorrect," since for example A might entail B, while B does not entail A. My response is that friends of the Law of Likelihood aren't committed to the ratio measure, and I see no reason why Bayesians must interpret the Law of Likelihood by using the F principle. Bayesians are under no obligation to force the Law of Likelihood onto that Procrustean bed.

Table 4.1 *Four conditional probabilities, each of the form Pr(Observation | Hypothesis)*

		Hypotheses	
		You have Covid-19	You do not have Covid-19
Observations	Test outcome is positive	w	x
	Test outcome is negative	y	z

is true constitutes a "crucial experiment" in the traditional sense of that term. It seems clear that O favors H_1 over H_2 in this example, and QUAL agrees, since the likelihoods are respectively 1 and 0.

For a less extreme bottom-up argument, suppose you take a reliable Covid-19 test and it comes out negative. The standard way to represent the reliability of a test procedure focuses on the likelihoods of hypotheses, as shown in Table 4.1. Suppose that $w + y = 1 = x + z$. Notice that y and x are error probabilities. Tests with smaller error probabilities are more reliable than tests with larger. A test procedure is at least minimally reliable precisely when $w > y$ and $z > x$.[9] Reliability has to do with "vertical" inequalities in the table, but these vertical inequalities entail inequalities that are horizontal: $w > x$ and $z > y$. A negative result of a reliable Covid-19 test favors the hypothesis that you don't have Covid-19 over the hypothesis that you do. A positive result has the opposite evidential significance. The Law of Likelihood captures what is going on in this example.

Where does the Law of Likelihood stand among nonBayesian statisticians? The situation is complicated. Friends of significance tests and of Neyman–Pearson hypothesis testing reject the law, sometimes arguing that the "optional stopping problem" reveals the law's deep flaws.[10] However, some nonBayesians accept the law. For example, when the Akaike information criterion (AIC) is used to evaluate two hypotheses, neither of which has any adjustable parameters, the one with the higher likelihood will

[9] Notice that the prior probability of your having Covid-19 plays no role in defining what it is for your test procedure to be reliable. A Covid-19 test procedure can have the same degree of reliability regardless of whether the disease is rare or rampant.

[10] See Sober (2008, pp. 72–78) for discussion.

always be the one with the better AIC score. This helps explain why Akaike (1973) regarded his AIC measure as an "extension of the maximum likelihood principle." More on this in Chapter 5.[11]

Thus far, I've discussed the justification of QUAL, but what about the quantitative law? Given QUAL, a quantitative measure of degree of favoring must depend only on the values of the two likelihoods, but why use the likelihood *ratio*? Why not use the likelihood *difference*? Suppose you take more than one Covid-19 test (always using the same test procedure, which is at least minimally reliable) and they all come back negative. It seems clear that the favoring that derives from these outcomes is stronger than the favoring that comes from the first outcome alone. QUANT affirms this judgment if the two test outcomes are probabilistically independent of each other, conditional on the hypothesis that you have Covid-19, and conditional on the hypothesis that you do not. In this case the likelihood ratio for the single negative test outcome is z/y and the ratio for n negative outcomes is z^n/y^n. If $z > y$, then $z^n/y^n > z/y$ when $n > 1$. The likelihood difference measure generates a different assessment. It says that $z-y$ is the degree to which the single negative test result favors the hypothesis that you lack Covid-19 over the hypothesis that you have Covid-19, and $z^n - y^n$ is the degree to which n negative test results favor no-Covid-19 over Covid-19. Since z and y are probabilities, $z^n - y^n$ shrinks towards zero as n increases.

There is more to be said about top-down and bottom-up assessments of the Law of Likelihood, but for the rest of this chapter I'll use this two-part principle without arguing for it further. I hope that the results obtained by applying the Law of Likelihood to the common and separate ancestry hypotheses will increase your confidence that QUAL and QUANT are sensible. I also hope that this enhanced confidence in the principles will lead you to reassess some of your intuitive judgments about the examples I'll discuss. This is precisely what I think needs to happen in connection with Darwin's claim that all adaptive similarities are almost useless as evidence for common ancestry. When bottom-up joins forces with top-down, you are using the methodology of *reciprocal illumination*, which is familiar in philosophy (Goodman 1965; Rawls 1970) and also in evolutionary biology (Hennig 1966).

[11] For another example, see Pawitan (2001), who argues that the Law of Likelihood can play an important role in nonBayesian statistics.

4.4 When Common Ancestry Has the Higher Likelihood

Although I introduced the idea of universal common ancestry in Figure 1.1, which is the phylogenetic tree from Darwin's *Origin*, I now want to discuss the somewhat simpler hypothesis that says that two or more objects have a common ancestor. The objects in question can be any entity that has an ancestry. I'll often take the objects to be taxa (e.g., biological species), but that is just for illustrative purposes.

Why think that the observed matching of two taxa has a higher probability if the common ancestry hypothesis is true than it would have if the hypothesis of separate ancestry were true? This inequality is not a consequence of the axioms of probability, so deriving it requires assumptions. Bear in mind that the assumptions I'll now describe *suffice* for the likelihood inequality; they are not *necessary*.[12] To describe these assumptions, I'll talk about the parameters x, a, y, b, and c shown in Figure 4.2, which I define as follows:

x = Pr(Humans have tailbones | C has a tailbone)
a = Pr(Humans have tailbones | C does not have a tailbone)
y = Pr(Monkeys have tailbones | C has a tailbone)
b = Pr(Monkeys have tailbones | C does not have a tailbonc)
c = Pr(C has a tailbone)

Here are the five assumptions:

(1) **Non-extreme probabilities:** $0 < x, a, y, b, c < 1$
(2) **Dependence on ancestors:** $(x - a)$ and $(y - b)$ are non-zero.
(3) **Cross-branch homogeneity:** $(x - a)$ and $(y - b)$ have the same sign.
(4) **Conditional independence:** An ancestor screens-off each of it descendants from the other, and it screens-off unrelated individuals from its descendants.
(5) **Cross-model homogeneity:** The probability of a descendant's having a tailbone does not depend on whether the common ancestry or the separate ancestry hypothesis is true.

I'll explain what "screening-off" means in a minute.

[12] This means that even if you doubt or reject the assumptions, the likelihood inequality may still obtain for different reasons.

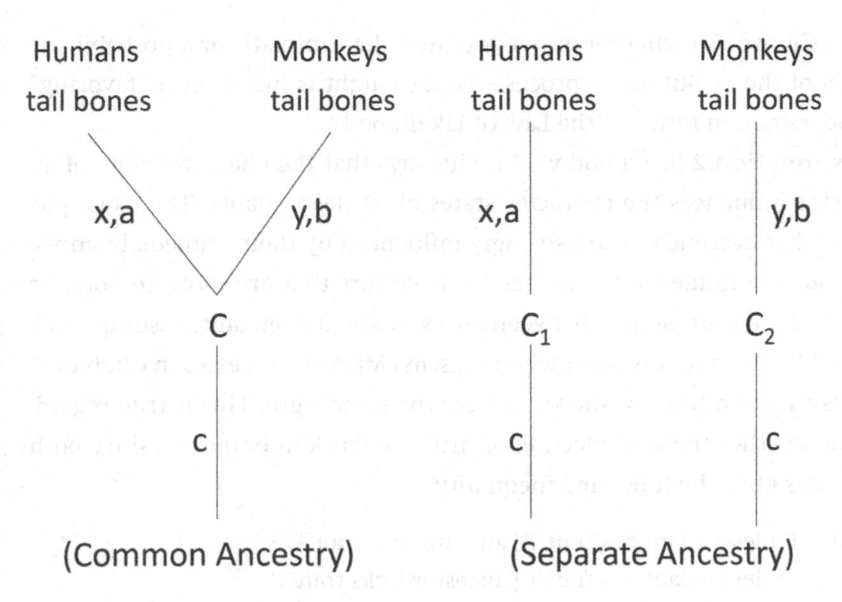

Figure 4.2 Parameterizing the common ancestry and the separate ancestry models.

Assumptions 4 and 5 entail that the likelihoods of common ancestry (CA) and separate ancestry (SA) are:

$$\Pr(\text{humans and monkeys both have tailbones} \mid \text{CA}) = cxy + (1-c)ab$$

$$\Pr(\text{humans and monkeys both have tailbones} \mid \text{SA}) = [cx + (1-c)a][cy + (1-c)b]$$

By using assumptions 1, 2, and 3, you can prove that the first of these likelihoods is greater than the second.[13] You don't need to know the point values of any of the five parameters to do this. The five assumptions are very general, and all the mathematical models for the evolution of a dichotomous trait that are currently used in phylogenetic biology obey those assumptions (Lemey et al. 2009). Here are some comments on each:

Assumption 1 (the probabilities that attach to branches are not extreme): What happens if you deny this assumption while retaining the others? For example, if $x = y = 1$ and $a = b = 0$ and $c = 1$ in Figure 4.2, the CA and the SA hypothesis have the same likelihood (= 1). The seemingly modest shift from non-extreme values to extreme values has a profound epistemological upshot. You might have thought that an observed matching

[13] **Exercise:** Use assumptions 1–3 to prove the inequality.

favors CA over SA whether or not you use a deterministic or a probabilistic model of the evolutionary process. That thought is incorrect if "favoring" is understood in terms of the Law of Likelihood.

Assumption 2 ($x \neq a$ and $y \neq b$): This says that the character state of an ancestor influences the character states of its descendants. The usual picture is that descendants are strongly influenced by their immediate ancestors, but the influence attenuates for ancestors that are more remote. As long as there is finite time between ancestor and descendant, Assumption 2 holds. I'll return to this point when I discuss Markov processes in Chapter 7.

Assumption 3 ($x - a$ and $y - b$ have the same sign): This is true regardless of whether there is selection or drift[14] at work in branches, since both processes obey the following inequality:

(COR) Pr(descendant has trait T | ancestor has trait T) >
Pr(descendant has trait T | ancestor lacks trait T)

This says that if a descendant has trait T, that outcome is more probable if its ancestor has T than it would be if its ancestor lacked T. In other words, ancestor and descendant are *positively correlated* with respect to whether they have trait T. COR is logically independent of the following inequality:

$$\Pr(\text{descendant has trait T} \mid \text{ancestor has trait T}) > \Pr(\text{descendant lacks trait T} \mid \text{ancestor has trait T})$$

This inequality entails that Pr(descendant has trait T | ancestor has trait T) > ½. It will be false if there is strong selection against trait T in the lineage connecting ancestor to descendant and there is sufficient time between the two.

Assumption 4 (ancestor C in Figure 4.2 screens-off each of its descendants from the other): This means that

$$\Pr(\text{humans have T \& monkeys have T} \mid \text{C has} \pm \text{T}) = \Pr(\text{humans have T} \mid \text{C has} \pm \text{T})\Pr(\text{monkeys have T} \mid \text{C has} \pm \text{T}).$$

"Screening-off" means conditional probabilistic independence. It can be defined equivalently as follows:

$$\Pr(\text{humans have T} \mid \text{C has} \pm \text{T}) = \Pr(\text{humans have T} \mid \text{C has} \pm \text{T \& monkeys have T})$$

[14] In this chapter, I use "drift" to mean that there is no variation in fitness with respect to trait T and its alternatives. That is, the trait is *strictly* neutral.

This second formulation conveys more clearly why the term "screening-off" is appropriate. To put the point informally, if you know that C has T, your probability that humans have T isn't changed by learning that monkeys have T. Ditto for knowing that C has notT. Screening-off may already be familiar to you in the context of Mendelian genetics. Suppose Jack and Jill are full siblings. Their genotypes are related to each other as follows:

$$\begin{aligned}&\Pr(\text{Jack has G1 \& Jill has G2} \mid \text{their parents have G3})\\&\quad = \Pr(\text{Jack has G1} \mid \text{their parents have G3}) \times\\&\qquad \Pr(\text{Jill has G2} \mid \text{their parents have G3}).\end{aligned}$$

The screening-off assumption applies to the SA hypothesis as follows:

$$\begin{aligned}&\Pr(\text{humans have T \& monkeys have T} \mid C_1 \text{ has} \pm \text{T \&} C_2 \text{ has} \pm \text{T})\\&\quad = \Pr(\text{humans have T} \mid C_1 \text{ has} \pm \text{T})\Pr(\text{monkeys have T} \mid C_2 \text{ has} \pm \text{T})\end{aligned}$$

Assumption 5 (the probability of a taxon's having a trait is independent of whether the common ancestry or the separate ancestry hypothesis is true): Here again, Mendelian genetics provides an intuitive analog. Your probability of having genotype G, given the genotypes of your parents, is not affected by whether you are an only child or have a sibling.

The upshot of these five assumptions is this: the probability that human beings and monkeys both have tailbones is greater if the two species have a common ancestor than it would be if the two species lacked a common ancestor. This is true even though (as Assumption 5 says) the probability that humans have a tailbone if there is a common ancestor is exactly the same as the probability that humans have a tailbone if there is no common ancestor. This isn't magic; the CA and SA hypotheses assign different probabilities to a *conjunction* though they assign the same probability to each *conjunct*.[15]

In summary, if species X and Y both have trait T and the five assumptions hold, the matching favors common ancestry over separate ancestry in the sense of the Law of Likelihood. Here trait T is dichotomous; a species either has T or it has notT. The assumptions can be modified in a natural way to apply to characters that have more than two states, and the

[15] **Question:** Is this property of the comparison of CA and SA hypotheses an instance of Simpson's paradox (Section 3.2)?

conclusion holds for them as well. In addition, the argument applies when more than two taxa are considered (Sober and Steel 2017).[16]

This likelihood result is worth considering in relation to Ockham's razor, the principle of parsimony. The CA hypothesis postulates one ancestor while the SA hypothesis postulates two. The CA requires that tailbones evolved at least once; the SA hypothesis requires that they evolved at least twice. The CA hypothesis is more parsimonious, but that is not a purely aesthetic feature that the hypothesis happens to have. The two hypotheses are identical in what they say (given the assumption of cross-model homogeneity), save for their difference with respect to the number of postulated causes. In this example, the more parsimonious hypothesis has the higher likelihood.

The likelihood inequality that follows from the five assumptions holds for specific versions of the CA and separate ancestry hypotheses. For example, notice that the common and separate ancestry hypotheses I've discussed do not say whether ancestors had tailbones. If you add to the CA hypothesis the postulate that C has a tailbone, and add to the separate ancestry hypothesis the postulate that C_1 and C_2 do too, the result is that the two hypotheses are *equal* in likelihood, given the five assumptions. The common and separate ancestry hypotheses I've analyzed are in this respect *bare bones*.

Another special feature of the likelihood analysis of common and separate ancestry is the assumption of cross-model homogeneity – the fact that parameters in the CA model are carried over to the separate cause model. This is plausible in the phylogenetic case, but in other assessments of common and separate cause hypotheses, it may not be. For example, if the students in a class are assigned an essay topic, and two of the students submit identical essays, the suspicion arises that they worked together (Salmon 1984). To assess this hypothesis we need to consider an alternative explanation – for example, that the students worked separately and independently. If you use different parameters in the common cause and the separate cause models, no likelihood inequality can be deduced. On the

[16] The argument just presented is inspired by Reichenbach's (1956) work on the principle of the common cause; for discussion of the specific connection to Reichenbach, see Sober (2015). Later in this chapter, I'll explain why I think that Reichenbach's principle is false.

other hand, if you abide by the assumption of cross-model homogeneity, you're committing to the idea that a student's probability of coming up with a given essay is the same whether the two students worked independently or worked together, and that might not be plausible.

4.5 Matches Sometimes Provide Evidence *Against* a Common Cause

As the example of the two students shows, comparing CA and SA hypotheses is a special case of the broader endeavor of comparing common and separate cause hypotheses. This means that there is a likelihood justification of the following thesis: When two objects have trait T, this matching favors the hypothesis that they have a common cause over the hypothesis that they do not. Although assumptions 1–5 make good sense in the context of testing common against separate ancestry in phylogenetics, there are examples in which the matching of two individuals fails to be evidence for the existence of a common cause. I described one of them in the previous section. Here are two more:

- Suppose there's a dominant gene that prevents an individual from having more than one offspring, and that Jack and Jill each have a copy of that gene. This matching favors the hypothesis that they are *unrelated* over the hypothesis that they are siblings.
- Suppose there are numerous businesses in Madison, each having one boss and several employees. Jack and Jill are employees and each is unhappy with his or her boss. Does that matching favor the hypothesis that they have the same boss over the hypothesis that they have different bosses? Not if Jack and Jill differ in the following respect: If Jack's boss is strict, that lowers the probability that he will be unhappy, whereas if Jill's boss is strict, that raises the probability that she will be unhappy. The fact that both are unhappy with their bosses now favors the hypothesis that they have *different* bosses.[17]

For an evolutionary example that resembles the second example, see Sober and Steel (2014).

[17] **Question:** In the two examples in this section, which of the five assumptions described in the previous section are violated?

4.6 Matches and Strength of Evidence

Can the likelihood approach throw light on which matches provide stronger evidence favoring CA over SA, and which provide weaker? I'll begin with a simple argument (Edwards 2007; Sober and Steel 2017). The likelihood ratio of the CA hypothesis to the SA hypothesis can be expanded as follows:

$$\frac{\Pr_{CA}(\text{X and Y have trait T})}{\Pr_{SA}(\text{X and Y have trait T})} = \frac{\Pr_{CA}(\text{X has trait T} \mid \text{Y has trait T})\Pr_{CA}(\text{Y has trait T})}{\Pr_{SA}(\text{X has trait T})\Pr_{SA}(\text{Y has trait T})}$$

If you use Assumption 5 (cross-model homogeneity), that

$$\Pr_{CA}(\text{Y has trait T}) = \Pr_{SA}(\text{Y has trait T}) = p$$

and assume further that the evolutionary process is uniform (meaning that simultaneous branches have the same probabilities of changing state), so that

$$\Pr_{CA}(\text{X has trait T}) = \Pr_{CA}(\text{Y has trait T}),$$

the likelihood ratio (LR^{CA}_{SA}) becomes

$$(*) \quad \frac{\Pr_{CA}(\text{X and Y have trait T})}{\Pr_{SA}(\text{X and Y have trait T})} = \frac{\Pr_{CA}(\text{X has trait T} \mid \text{Y has trait T})}{\text{p}}.$$

Suppose, finally, that if X and Y have a common ancestor, then the amount of time between X and Y and their most recent common ancestor is very small. This entails that the likelihood ratio is approximately $\frac{1}{p}$.

Given that the likelihood ratio gets bigger as p gets smaller, there is a simple argument that lends credence to Darwin's thesis that adaptive similarities provide little evidence for CA. The argument does not describe the absolute amount of evidence that adaptive similarities provide. Rather, it involves a comparison: if the value of p when T is adaptive is greater than the value of p when T is neutral, which in turn is greater than the value of p when T is deleterious, then deleterious similarities provide stronger evidence for CA than neutral similarities do, and neutral similarities provide stronger evidence for CA than adaptive similarities do.[18]

The $\frac{1}{p}$ argument is a good starting point, but it has two limitations. The first is that the argument is formulated for just two taxa; Sober and Steel (2017) generalize the result. The second limitation is the assumption that

[18] I'll explain in what follows that there are counterexamples to this second conclusion.

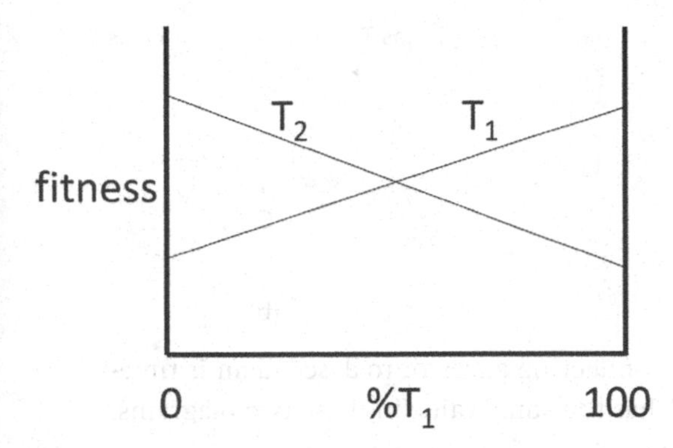

Figure 4.3 Selection for the majority trait.

if X and Y have a common ancestor, then they have a very recent common ancestor. This limitation can be lifted by considering a stationary equal input model.[19] In this model the likelihood ratio LR_{SA}^{CA} is greater than unity when the temporal separation of A and B and their most recent common ancestor is finite (though it asymptotically approaches unity as temporal separation is made large) and the ratio is larger the smaller p is (see Sober and Steel 2017 for details).

The mathematical analysis I've described so far reinforces Darwin's judgments about stronger versus weaker evidence for CA, but now it's time for a surprise. There is a kind of natural selection that allows adaptive similarities to furnish stronger evidence for CA than neutral similarities are able to provide; it is depicted in Figure 4.3. There is frequency dependent selection in this example, but it differs from the example of frequency dependence that I discussed in connection with altruism and selfishness in Chapter 3. In Figure 4.3, there is selection for the majority trait. In this situation, the population will probably evolve to either 100 percent T_1 if it begins with T_1 in the majority, and it will probably evolve to 100 percent T_2 if it begins with T_2 in the majority. There are two *adaptive peaks* in this model and the population ascends to one of them or the other.[20] The model allows that a population might evolve from 100 percent of one trait to 100 percent of the other, but that is pretty improbable, as the population will need to swim against the tide of natural selection in the first part of its evolution.

[19] An equal input model for an n-state character X is one in which Pr(descendant has X_i | ancestor has X_j) = Pr(descendant has X_i | ancestor has X_k) for all $j,k \neq i$. Dichotomous traits automatically satisfy *the* equal-input assumption.

[20] **Question:** What does "adaptive peak" mean here? Hint: think about w-bar.

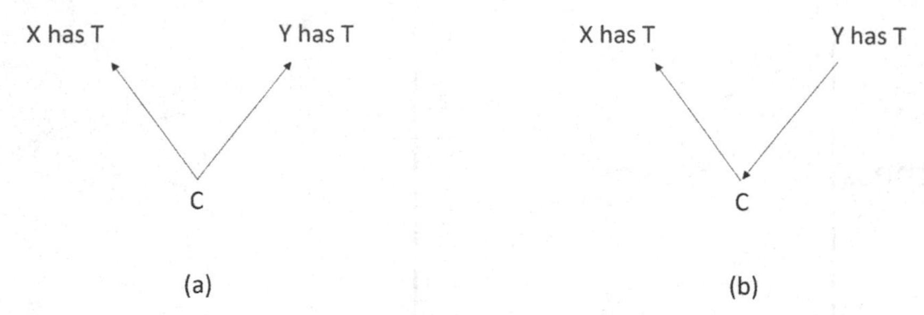

Figure 4.4 When the process connecting ancestor to descendant is time-reversible, Pr(X has T | Y has T) has the same value in these two diagrams.

The evolutionary process shown in Figure 4.3 is *time-reversible*, meaning that in population Z,

$$\Pr(\text{Z is 100\% T at time } t_2 \mid \text{Z is 0\% T at } t_1) = \Pr(\text{Z is 0\% T at time } t_2 \mid \text{Z is 100\% T at } t_1)$$ [21]

Neutral evolution is also time-reversible. This property of the two processes comes in handy in connection with the question of how the difference between neutral evolution and frequency dependent selection for the majority trait affects the likelihood ratio of CA and SA. The handiness is represented in Figure 4.4.

In Figure 4.4b, $\Pr_{CA}$(X has trait T | Y has trait T) has a higher value if there is selection for the majority trait than it would have if there were neutral evolution.[22] Suppose the value for *p* is the same for the two processes. This means that $\frac{\Pr(\text{X has T} \mid \text{Y has T})}{p}$ is bigger when there's frequency dependent selection for the majority trait than it would be if there were drift. Plug this fact into the asterisk proposition above and you'll see that $\frac{\Pr_{CA}(\text{X and Y have trait T})}{\Pr_{SA}(\text{X and Y have trait T})}$ is greater when there is selection for the

[21] **Question:** Is frequency independent selection time-reversible?

[22] Here's why. If evolution is neutral, let's suppose that the probability of a lineage's evolving from 100% T to 50% T is q and its probability of evolving from 50% T to 0% T is q as well. So the probability of going from 100% to 0% if there is drift is q^2. In contrast, when there's selection for the majority trait, the probability of going from 100% to 50% is $q - c$ (where c is positive) and its probability of going from 50% to 0% is $q + c$. This means that the probability of going from 100% to 0% in this case is $(q - c)(q + c) = q^2 - c^2$. Change is more probable when there is drift, so stasis is more probable when there is frequency dependent selection for the majority trait.

majority trait than it would be if there were neutral evolution. This shows that adaptive similarities sometimes favor CA over SA more than neutral similarities are able to do.

This result about selection for the majority trait throws light on the example of the near universality of the genetic code, which I mentioned as an apparent counterexample to Darwin's claim that adaptive similarities are "nearly valueless" as evidence for CA. The genetic code is often described as "arbitrary," meaning that numerous alternative codes would have worked. However, this point about arbitrariness leaves open the possibility that the genetic code used in a species is subject to natural selection. In a sexual population, there is powerful selection for having the genetic code that other members of the species possess. An organism that fails to conform has almost no chance of producing viable fertile offspring. A similar point pertains to species in which reproduction is exclusively asexual if a change in an organism's genetic code has a high probability of rendering it inviable. In Figure 4.3, traits T_1 and T_2 might be said to be "arbitrary," in that the organisms in a population that is 100 percent T_1 have the same fitness as the organisms in a population that is 100 percent T_2. However, the figure can be modified so that T_1 is optimal, meaning that T_1's fitness when T_1 is at 100 percent exceeds T_2's fitness when T_2 is at 100 percent. This situation is relevant in light of the argument that the near-universal code that biologists observe is optimal (Freeland et al. 2000); this optimality does not undercut the idea that the shared code provides substantial evidence for CA.[23] On the other hand, if there were only one viable code that is physically possible, the universality of that code would not favor CA over separate.

Figure 4.5 summarizes how the matching of taxa X and Y with respect to trait T affects the likelihood ratio of the CA and SA hypotheses. A partial ordering is depicted here, in that nothing is said about how selection against trait T compares with selection for the majority trait.[24]

[23] However, note that that the time-reversibility argument I made in connection with Figure 4.4 does not apply if the near-universal code is optimal.

[24] **Question:** Do the ideas presented in this section help solve the puzzle described at the start of the chapter concerning Darwin's discussion of the "framework of bones"?

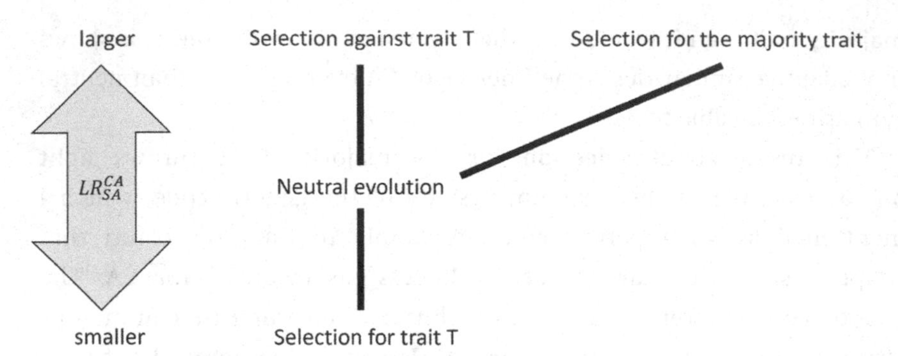

Figure 4.5 Given that taxa X and Y both have trait T, the likelihood ratio of CA to SA $\left(LR_{SA}^{CA}\right)$ depends on the evolutionary process at work.

4.7 Hypothetico-Deductivism

It is often said that testing a scientific theory requires one to determine what the theory predicts about observations. The results described in Figure 4.5 cast doubt on that "truism." I take it to be part of the concept of prediction that the following is true: If hypothesis H predicts that observation statement O is true, then Pr(O | H) > ½. Hypotheses don't predict observational outcomes that they say will probably not occur. However, when you consider how strongly an observed matching of two taxa favors CA over SA, it matters not at all whether the competing hypotheses predict the observations in this sense. What matters is the value of the likelihood ratio, and *small likelihoods can give rise to big likelihood ratios.*[25] The idea that theories can be tested empirically only by seeing what they predict is a vestige of a bygone philosophical age in which deductive logic provided the only tools for characterizing how theories are related to observations. The name for that outmoded idea is *hypothetico-deductivism*.

[25] Here's another example that illustrates this point. Suppose two individuals match with respect to 10 rare genes (Sober 2008, p. 52). If each gene has a frequency of $\frac{1}{1{,}000}$ in the population, then the probability that the individuals match with respect to all ten is $\left[\left(\frac{1}{1{,}000}\right)\left(\frac{1}{2}\right)\right]^{10}$ if they are siblings, but it's $\left[\left(\frac{1}{1{,}000}\right)\left(\frac{1}{1{,}000}\right)\right]^{10}$ if they are unrelated. Both likelihoods are tiny, but the ratio of the first to the second is $(500)^{10}$, which is gigantic.

4.8 Mismatches

I've used the Law of Likelihood to discuss the evidential significance of "matches." Now it's time to consider what the law says about mismatches. My starting point is the fact that the Law of Likelihood has a symmetry property:

O favors H_1 over H_2 if and only if notO favors H_2 over H_1.

This is a consequence of the Law of Likelihood because it is a theorem of probability theory that

$$\Pr(O \mid H_1) > \Pr(O \mid H_2) \text{ if and only if } \Pr(\text{not}O \mid H_1) < \Pr(\text{not}O \mid H_2).$$ [26]

Thus, if two species' having trait T would favor CA over SA, and their both having notT would do the same, then their mismatching (wherein one has T and the other has notT) must favor SA over CA. Considering the numerous differences that exist between humans and monkeys, do these numerous mismatches "swamp" the traits on which the two taxa match, with the result that the total evidence favors SA over CA?

To explore this question, let's begin by considering the two columns in Table 4.2 that represent the likelihoods of CA and SA, relative to different possible observations. For simplicity, I'm assuming that the lineage leading to taxon X and the lineage leading to taxon Y have the same probabilities of changing state. Given assumptions 1 – 5, c_1 and c_3 are positive.[27] The likelihoods in the second column automatically sum to one. The likelihoods in the first column must do so as well, so $c_1 - 2c_2 + c_3 = 0$. Suppose you observe that X and Y both have trait T_1 and that X has T_2 while Y does not. These two observations disagree about which hypothesis has the higher likelihood. How should you interpret their joint testimony? If the two traits conform to the probabilities shown in the above table, and if the two observations are probabilistically independent of each other, conditional on CA and conditional on SA, you can write:

$$\frac{\Pr(\text{X and Y have } T_1 \text{ \& X has } T_2 \text{ and Y has not} T_2 \mid CA)}{\Pr(\text{X and Y have } T_1 \text{ \& X has } T_2 \text{ and Y has not} T_2 \mid SA)}$$

$$= \left(1 + \frac{c_1}{p^2}\right)\left(1 - \frac{c_2}{p(1-p)}\right)$$

[26] **Exercise:** Prove this theorem.

[27] Notice that the likelihood ratio in the first row depends on c_1 as well as on p. This helps clarifies why the $\frac{1}{p}$ argument described in the previous section provides an incomplete picture.

Table 4.2 *The likelihoods of common ancestor (CA) and separate ancestor (SA) relative to four possible observations, and the four resulting likelihood ratios*

		Pr(Observation | CA)	Pr(Observation | SA)	LR_{SA}^{CA}
Observations	X has T & Y has T	$p^2 + c_1$	p^2	$\frac{Pr(X \text{ and } Y \text{ have } T \mid CA)}{Pr(X \text{ and } Y \text{ have } T \mid SA)} = 1 + \frac{c_1}{p^2}$
	X has T & Y has notT	$p(1-p) - c_2$	$p(1-p)$	$\frac{Pr(X \text{ has } T \text{ and } Y \text{ has notT} \mid CA)}{Pr(X \text{ has } T \text{ and } Y \text{ has notT} \mid SA)} = 1 - \frac{c_2}{p(1-p)}$
	X has notT & Y has T	$p(1-p) - c_2$	$p(1-p)$	$\frac{Pr(X \text{ has notT and } Y \text{ has } T \mid CA)}{Pr(X \text{ has notT and } Y \text{ has } T \mid SA)} = 1 - \frac{c_2}{p(1-p)}$
	X has notT & Y has notT	$(1-p)^2 + c_3$	$(1-p)^2$	$\frac{Pr(X \text{ and } Y \text{ have notT} \mid CA)}{Pr(X \text{ and } Y \text{ have notT} \mid SA)} = 1 + \frac{c_3}{(1-p)^2}$

The total evidence favors CA over separate precisely when

$$\left(1+\frac{c_1}{p^2}\right)\left(1-\frac{c_2}{p(1-p)}\right)>1,$$

which is true if and only if

$$\frac{c_1}{p^2}>\frac{c_2}{p(1-p)}+\frac{c_1c_2}{p^3(1-p)}.$$

This condition holds for some values of c_1, c_2, and p, but not for others. It is understandable that friends of CA hypotheses might hope that a match always "trumps" a mismatch in terms of their opposing impact on the competition between the CA and the SA hypotheses. The simple formulation of the problem just explored provides no such blanket assurance.

4.9 Correlations to the Rescue?

Can the puzzle described in the previous section – of how to compare matches and mismatches for their epistemic significance – be set to one side? Instead of evaluating the evidential meanings of traits one by one and then comparing them, maybe you can step back from these details and focus on the question of whether the observed traits of taxa X and Y are *correlated*. If there is a positive correlation, you might then invoke a principle that Reichenbach (1956) called the *principle of the common cause*:

> If A and B are positively correlated, then either A causes B, B causes A, or the two have a common cause.

That label makes sense because the principle of the common cause entails that

> (PCC) If A and B are positively correlated, and neither causes the other, then A and B have a common cause.

If taxa X and Y exist at the same time, their traits are positively correlated, and if cause must precede effect, then the PCC provides a simple deductive argument for CA.[28]

PCC is central to current methodologies of causal modeling (Glymour, Spirtes, and Scheines 1993; Pearl 2000; Hitchcock and Redei 2021), but

[28] **Question:** What are the similarities and differences between the PCC and the likelihood comparison of common ancestry and separate ancestry?

the trouble is that it is false. Philosophers have sometimes cited quantum mechanics as the source of counterexamples (Van Fraassen 1980), but numerous counterexamples exist in less esoteric domains. Udny Yule (1926) pointed this out in connection with time-series data. I described a hypothetical example of this type involving the observation that sea levels in Venice and bread prices in Britain both steadily increased over the past two centuries (Sober 2001b). Years with higher-than-average bread prices in Britain tend to have higher-than-average sea levels in Venice, but this could easily be because two separate and causally unconnected processes made increase more probable than decline in each time step.[29]

It isn't enough for my purposes that the *data* on Venice and Britain have the pattern just described. What matters is how the *underlying probabilities* behave. For a simple illustration, let's divide the 200 years into 200 × 365 = 73,000 days ($t_1, t_2, \ldots, t_{73000}$). Suppose you have two data sets, one for Venice, the other for Britain; on each of 1,000 randomly selected days, you record the Venetian sea level and the British bread price. Imagine a graph of these data points, with time on the x-axis and sea levels and bread prices on the y-axis. You then do a linear regression on each data set and find the best-fitting straight line for Venice and the best-fitting straight line for Britain. Each line is closer to its 1,000 data points than any competing straight line (it has maximum likelihood), but it does not go through those data points exactly. The straight lines represent the inferred *expected values* for bread prices and sea levels on each of those 73,000 days. Suppose that these two straight lines are true descriptions *of the underlying processes* – each day, the expected price of bread goes up, and so does the expected sea level. The two regression lines entail this correlation:

$$\Pr(\text{sea level at t is higher than average \& bread price at t is higher than average}) > \Pr(\text{sea level at t is higher than average}) \times \Pr(\text{bread price at t is higher than average}).$$

Roughly speaking, this inequality holds because h > h^2, for any h that is strictly between 0 and 1.[30]

[29] The point here is not restricted to cases in which there is monotonic increase. Suppose each of two causally unconnected processes leads a quantity to first increase and then decline, or suppose that each induces a sinusoidal trajectory.

[30] **Exercise**: There is nothing special about *time* series in this and similar examples. To see why, construct a *spatial* analog of the Venice-Britain example.

Table 4.3 *The result of tossing two coins three times each*

	toss 1	toss 2	toss 3
coin 1	H	H	T
coin 2	H	H	T

The term "spurious correlation" comes to mind here, but unfortunately it is used in different ways. Sometimes the correlation of two events is said to be spurious if neither causes the other. That seems odd; siblings are correlated in the genes they have because they have the same parents; the correlation is real, not spurious. Sometimes correlations are said to be spurious because they violate the PCC. This strikes me as a poor choice of terminology. It's the principle of the common cause that is spurious, not the genuine probabilistic correlation of Venice and Britain. I suggest that a spurious correlation is just a frequency *association* in the data that sits side-by-side with underlying probabilistic *independence*.[31] Evidence for there being a genuine correlation between Venice and Britain comes from the fact that sampling a new set of 1,000 days will produce the same pattern in the data.[32]

A simple coin-tossing example illustrates this conception of what a spurious correlation is. Suppose you toss two coins three times each and get the data shown in Table 4.3. There is an association in your data, in that

[31] Tyler Vigen (2015) used data mining to assemble a treasure trove of hilarious examples in his book *Spurious Correlations*. A few of his examples are on-line at www.tylervigen.com/spurious-correlations. **Question:** Is Vigen's graph relating worldwide non-commercial space launches and sociology doctorates awarded in the United States a counterexample to the principle of the common cause?

[32] Although the Venice-Britain example is a counterexample to Reichenbach's PCC, it is not a counterexample to the Causal Markov condition, which also is a mainstay of causal modeling. This states that $Pr(E = j \mid C = i) = Pr(E = j \mid C = i \ \& \ N = k)$, where C is the (nonempty) set of direct causes of E, and N is anything distinct from C and E that is not caused by E. Here C, E, and N are variables. Notice that this equality describes a screening-off relation. In the example at hand, it is true, for any $t_2 > t_1$, that $Pr[V(t_2) = y \mid V(t_1) = x] = Pr[V(t_2) = y \mid V(t_1) = x \ \& \ B(t) = z]$, where V is the Venetian sea level, B is the British bread price, and t is any time you please. Hausman and Woodward (1999) combine the Causal Markov condition just described and Reichenbach's PCC into a single principle, which they call the Causal Markov condition. I prefer to keep the two principles separate.

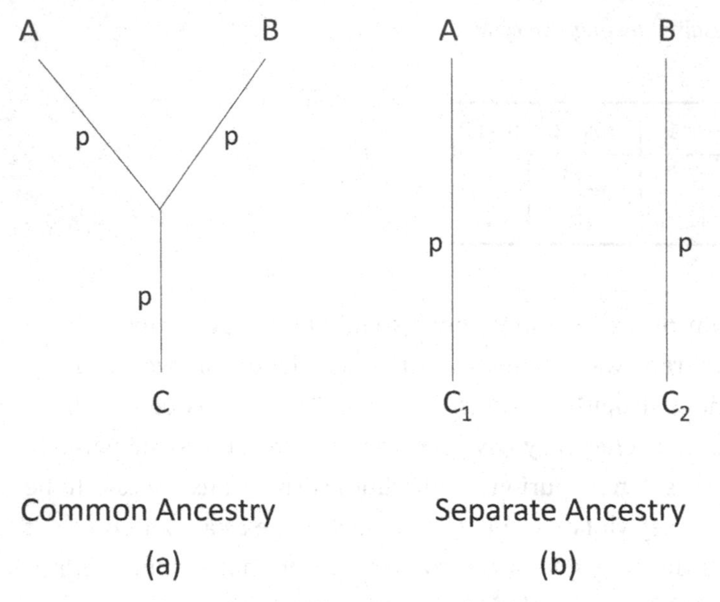

Figure 4.6 A parameterization of the common ancestry and separate ancestry hypotheses that violates the assumption of cross-model homogeneity

freq(coin 1 lands heads & coin 2 lands heads) = 2/3, which is greater than freq(coin 1 lands heads)freq(coin 2 lands heads). It is entirely possible that your data are misleading, however, since the underlying reality might be that the coins are probabilistically independent of each other. You can check whether they are by doing a new experiment. If new data reveal that the coins are independent of each other, the suggestion that they are correlated is spurious. It is spurious because it is false. The *association* in your first data set isn't spurious; it is real. The trouble is that the observed association is misleading. Perhaps the term "spurious correlation" should be replaced with "misleading association."

Setting Venice and Britain aside, I want to describe another reason to think that simultaneous events can be correlated without tracing back to a common cause. Consider the CA and SA hypotheses depicted in Figure 4.6. Unlike Figure 4.2, which is about the evidential meaning of a single trait (tailbones), Figure 4.6 describes numerous dichotomous traits (each of whose states will be coded as "0" or "1"). Suppose that traits evolve in both the CA and the SA hypothesis by following a symmetric rule of character evolution, where each branch is such that p = Pr(End = 1 | Start = 0) = Pr(End = 0 | Start = 1) = ¼. Notice that the placement of "p" in the figure violates assumption 5 of cross-model homogeneity. In the common cause model (Figure 4.6a), suppose that C = 0 is the ancestral state of each trait. Then the four possible combinations of character states for A and B on each character have these probabilities:

$$Pr_{CA}(A = 1 \& B = 1) = 3/16$$
$$Pr_{CA}(A = 1 \& B = 0) = 3/16$$
$$Pr_{CA}(A = 0 \& B = 1) = 3/16$$
$$Pr_{CA}(A = 0 \& B = 0) = 7/16$$

It follows that $Pr_{CA}(A = 1) = Pr_{CA}(B = 1) = 3/8$, so A and B are correlated, since

$$Pr_{CA}(A = 1 \& B = 1) > Pr_{CA}(A = 1)Pr_{CA}(B = 1)$$

Similarly, it follows that $Pr_{CA}(A = 0) = Pr_{CA}(B = 0) = 10/16$, with the result that

$$Pr_{CA}(A = 0 \& B = 0) > Pr_{CA}(A = 0)Pr_{CA}(B = 0).$$

Just as one would expect, the common cause arrangement entails a positive correlation between the two effects.

By violating the assumption of cross-model homogeneity, you can construct a SA hypothesis (Figure 4.6b) that gives rise to exactly the same correlation. Suppose there are two types of trait that evolve in that genealogy. Type-1 traits all have C_1 and C_2 in state 1, whereas type-2 traits all have C_1 and C_2 in state 0. Suppose further that the frequency of type-1 traits is 25 percent and the frequency of type-2 traits is 75 percent. If the probability of change on each branch in the SA genealogy is $p = 1/4$, the hypothesis of SA and the CA hypothesis entail the same correlation.[33]

Friends of Reichenbach's PCC may reply that this separate cause model is far-fetched. That may be, but the point is that Reichenbach's principle entails that such models *can never be true*. What's the argument for this claim? Friends of the PCC will probably want to claim that the correlation between C_1 and C_2 must be due to their having a common cause, and once that is added to the Figure 4.6b model, the problem I've described disappears. I agree that there *may* be a common cause in this situation, but Reichenbach's principle goes further; it says that there *must* be one. This is where I get off the bus.[34]

[33] **Question:** Is this example an instance of Simpson's paradox (Section 3.2)?

[34] This distinction between *may* and *must* also applies to the Venice–Britain example.

This example, far from being one-off, is an instance of a general pattern. For any common cause model that obeys assumptions 1–5 and whose probabilistic parameters are assigned values, there exists a separate cause model whose parameters can be assigned values such that the two models entail the very same correlation between effects. Sober and Steel (2002) derive this result, calling it *the mimic theorem*. You may be tempted to complain that the mimicking separate cause model is *ad hoc* or is parasitic on the common cause model. I agree, but this comment concerns the activities of model-builders (their motivation and how they constructed the model); it does not impugn the model itself. The fact remains that the common cause and separate cause models just described entail the same correlation.

The mimic theorem is not a cause for despair; rather, it points to a valuable lesson. Mimicry is possible when there are no constraints on how the common and separate cause models are constructed. The way forward is to focus on models that are constrained. The assumption of cross-model homogeneity is such a constraint; this assumption is reasonable in the context of inferences about common ancestry and separate ancestry, and in phylogenetic inference more generally (a topic I'll discuss in Chapter 7).

The mimic theorem is an example of a standard problem in philosophy of science – the underdetermination of theory by evidence. Underdetermination means that for any finite data set, there are alternative incompatible hypotheses that fit the data equally well. Philosophers sometimes illustrate this problem by inventing "weird" postulates – Descartes' evil demon and Reichenbach's (1958) universal forces are two examples. In contrast, the mimic theorem applies to competing hypotheses that use exactly the same concepts.

I took up the topic of correlation because of a problem – the problem of assessing the likelihoods of a CA and a SA hypothesis when your data set includes both matches and mismatches. The point thus far is not that correlations are worthless, but that their usefulness should not be overstated. To find a way forward, I'll describe a simple example used by Reichenbach (1956, p. 158). Suppose you observe two geysers (G1 and G2) over several years in the same valley and notice that they each erupt for an hour each week, but when one erupts, the other almost always does. This means that the frequency with which they both erupt is greater than the product of the frequencies with which each erupts. There is an *association* in your data set, meaning that

$$\text{freq}(\text{G1 and G2 erupt at time t}) > \text{freq}(\text{G1 erupts at t})\text{freq}(\text{G2 erupts at t}).$$

Now let's consider two probability models that might explain this observation. These models can be applied to the observed association, but they are more general, in that they make predictions about other data sets that might be drawn from the two geysers.

A common-cause probabilistic model that obeys assumptions 1–5 entails that the two geysers are correlated:

$$\Pr(\text{G1 and G2 erupt at time t}) > \Pr(\text{G1 erupts at t})\Pr(\text{G2 erupts at t}).$$

$$\Pr(\text{neither G1 nor G2 erupts at time t}) > \Pr(\text{G1 fails to erupt at t}) \times \Pr(\text{G2 fails to erupt at t}).$$

This description of the positive correlations entailed by the common cause hypothesis might seem to be fixated on matches, but mismatches are covered by the two correlation claims, since they and the following conjunction are equivalent:

$$\Pr(\text{G1 erupts at t and G2 does not erupt at t}) < \Pr(\text{G1 erupts at t}) \times \Pr(\text{G2 does not erupt at t}).$$

$$\Pr(\text{G1 does not erupt at t and G2 erupts at t}) < \Pr(\text{G1 does not erupt at t}) \times \Pr(\text{G2 erupts at t}).$$

A competing separate cause hypothesis (also conforming to the five assumptions) says that the eruptions of the two geysers are probabilistically independent, which can be described (redundantly) by changing the above four inequalities into equalities.

This example shows how an observation can favor a common cause over a separate cause hypothesis where the observation that does the favoring is an *association*, not the raw data that describes what happened hour by hour. This analysis does not have the defect of being one-sided – of attending to matches and ignoring mismatches. However, there is an important feature of the geyser example that points to a new question about using associations to test CA against SA.

In Reichenbach's geyser example, each observed eruption has the same evidential meaning as every other, but in the comparison of common and SA hypotheses, it is entirely possible that different traits have different evidential meanings; think of the results in Section 4.6, which describe how the evidential meaning of a match depends on the evolutionary process at work. One way forward is to restrict one's data to traits that follow the same rules in their evolution; for example, you might want to restrict your attention to traits that evolve by neutral evolution. This type of evolutionary process is the subject of the next chapter.

4.10 Concluding Comments

The first epistemological question raised in this chapter is qualitative:

- If species X and Y both have trait T, does this matching favor the hypothesis that they have a common ancestor over the hypothesis that they do not?

The second is quantitative:

- If species X and Y both have trait T, how strongly does this matching favor the hypothesis that they have a common ancestor over the hypothesis that they do not?

I briefly discussed the justification of the Law of Likelihood, and then used the qualitative formulation of the law to address the first question and the quantitative formulation to address the second.

For both questions, the answers I described depend on assumptions about the evolutionary processes at work in the lineages leading to X and to Y. With respect to the first question, assumptions 1–5 suffice to show that the observation that X and Y both have trait T favors common ancestry over separate ancestry. With respect to the second, those five assumptions, plus models of different evolutionary processes, together suffice for the conclusions drawn concerning how the character of the evolutionary process affects how strongly an observed matching favors common over separate ancestry.[35]

I then turned to mismatches. Darwin was right that different taxa have "much in common," but those similarities coexist with numerous differences. An assessment of the common ancestry hypothesis should take account of *all* the evidence, both the evidence *for* and the evidence *against*. That said, the result presented in Section 4.8 that compares a match with

[35] By focusing on matches as evidence for common ancestry, I have ignored quantitative characters or discrete n-state characters that are ordered; in both, "near-matches" should count as evidence for common ancestry and "large-mismatches" should count as evidence against. For discussion, see Sober (2008, pp. 305–306) and Helgeson (2018).

a mismatch may be a bit disappointing – whether a match "trumps" a mismatch depends on further probabilistic details. This led me to consider abandoning the assessment of traits one by one and to see instead whether correlations between taxa can be parlayed into an argument for common ancestry.

After describing Reichenbach's Principle of the Common Cause, I presented objections to that principle. The PCC is too strong, but Reichenbachian ideas throw light on how a data set containing both matches and mismatches can be assessed for its evidential bearing on the common ancestry and the separate ancestry hypotheses. If the traits in your data set evolve independently of each other and follow the same rules of evolution, then an observed association will favor common ancestry over separate ancestry; this observed association takes account of both matches and mismatches.

I have focused in this chapter on the epistemology of common ancestry, but there is another type of genealogical hypothesis that is philosophically interesting. To describe it, I need to move from the convenient example of two organisms or species to three. If human beings, chimpanzees, and gorillas have a common ancestor; the new question is whether human beings and chimpanzees are more closely related to each other than either is to gorillas. To test this hypothesis, we need to contrast it with alternatives; (HC)G needs to tested against H(CG) and (HG)C. This type of phylogenetic inference problem will be explored in Chapter 7.

5 Drift

When Darwin (1859, p. 6) said in the *Origin* that natural selection "has been the main but not the exclusive cause" of evolution, what other causes did he have in mind? I described one of them in Section 1.6 – Darwin's endorsement of "use and disuse," the Lamarckian idea that traits acquired in the lifetime of a parent can be inherited by offspring – noting that this idea fell by the way in the Modern Synthesis. A second alternative to natural selection that Darwin identified turned out to have a brighter future:

> Variations neither useful nor injurious would not be affected by natural selection, and would be left either a fluctuating element, as perhaps we see in certain polymorphic species, or would ultimately become fixed, owing to the nature of the organism and the nature of the condition (Darwin 1859, p. 81).[1]

Darwin's nod to what is now called "neutral evolution" or "random genetic drift" or just "drift" is well taken, but the idea didn't play much of a role in his theorizing. That turn of events occurred in the second half of the twentieth century when innovative mathematical theories, coupled with increasingly fine-grained data sets on genetic variation, brought the neutral theory to center stage.[2]

To ease our way into this technical subject, I'll begin by discussing a paper that appeared in Darwin's own lifetime. It isn't about genes; indeed, it's on a subject that seems at first glance to be worlds away from evolutionary biology.

[1] See Gigerenzer et al. (1989, pp. 132–141) for discussion of what Darwin meant by "chance."

[2] See Millstein (2021) and Abrams (2007) for historical and conceptual discussion of the different meanings that "genetic drift" has been assigned.

5.1 The Extinction of Family Names

Francis Galton, who was Darwin's cousin, is famous for inventing numerous basic concepts in statistics (e.g., correlation and variance). In 1873, he published a mathematical problem about the disappearance of family names, inviting readers to submit solutions. Henry Watson replied, after which he and Galton wrote an article called "On the probability of the extinction of families" (Watson and Galton 1875). The mathematical problem that Galton posed was highly structured; it concerns only one of two possible explanations of an observation:

> The instances are very numerous in which surnames that were once common have since become scarce or have wholly disappeared. The tendency is universal, and, in explanation of it, the conclusion has been hastily drawn that a rise in physical comfort and intellectual capacity is necessarily accompanied by diminution in "fertility," using that phrase in its widest sense and reckoning abstinence from marriage as sterility. If that conclusion be true, our population is chiefly maintained through the "proletariat," and thus a large element of degradation is inseparably connected with those other elements which tend to ameliorate the race. On the other hand, M. Alphonse De Candolle has directed attention to the fact that, by the ordinary law of chances, a large proportion of families are continually dying out … and it evidently follows that, until we know what that proportion is, we cannot estimate whether any observed diminution of surnames among the families whose history we can trace, is or is not a sign of their diminished "fertility."

Watson and Galton's first hypothesis says that intelligent people who lead comfortable lives have fewer babies than members of the "proletariat."[3] The authors say that this hypothesis has been "hastily drawn" and then set it aside, focusing instead on the second hypothesis, which says that there are no differences in fertility.

Watson and Galton were right that the first hypothesis was "hastily drawn," but it is curious that they say this. In his 1869 book *Hereditary Genius*, Galton argued that intelligence runs in families because it is

[3] Watson and Galton say that this first hypothesis, if true, will lead to "degradation" (presumably of health and intelligence) and that those who seek to ameliorate the life of the poor are accelerating that degradation. The authors are echoing Malthus (Section 2.1).

biologically inherited.[4] In the data set that Galton assembled in support of this hypothesis, the sons of "eminent" fathers are eminent themselves more frequently than the sons of men who are not eminent.[5] In his 1883 book, *Inquiries into Human Faculty and its Development*, Galton used his conclusion that health and intelligence are inherited as his main argument for *eugenics* (a word he coined), urging governments to encourage people who are intelligent and healthy to have more babies. Such incentives fell under the heading of "positive eugenics." Eugenics became a powerful political program in many countries and found supporters across the political spectrum. Side-by-side with this carrot were multiple brutal sticks that implemented "negative eugenics," where governments compelled people judged to be mentally or physically impaired to be sterilized.[6] The United States was a prime developer of such policies, and the US example prompted other countries to do the same (Paul 1995; Kevles 1998; Levine 2017). In the United States, eugenic concerns led to laws prohibiting "miscegenation" and to immigration policies that tightly limited immigration from outside northern Europe. Adolf Hitler took negative eugenics a big step further – to the murdering of millions of Jews, Roma, Slavs, homosexuals, and others deemed "unfit." Well before the Second World War, R. A. Fisher, a founding father of twentieth-century evolutionary theory and of statistics as well, devoted the second half of his *Genetical Theory of Natural Selection* (Fisher 1930) to defending eugenics. The first half of that book is widely regarded as a masterpiece. The second is widely regarded as an embarrassment. Fisher's worry, like Galton's, was the decline of civilization.

What is now called the Galton–Watson model for explaining the extinction of family names *denies* that people with different names differ in their fertility. Names disappear merely because of "the ordinary law of chances." For example, suppose that each of the males in a population has a different

[4] It now is standard to recognize that a trait can run in families without being caused by genes. Spoken languages are an example.

[5] **Question:** How does the concept of positive correlation (Section 3.2) apply to Galton's finding?

[6] Negative eugenics existed de facto well before Galton invented the name. Physicians sometimes took it upon themselves to sterilize individuals without their consent in the United States, their victims typically being black people, indigenous people, poor people, people thought to be mentally impaired or to have "bad morals," and people with syphilis or epilepsy (Reilly 1991; Peterson forthcoming). It Goverment-guided eugenics took hold only later.

surname, that surnames are always inherited from one's father, and that people don't change their surnames. In the Galton–Watson model, each male is characterized by the same probability distribution; each has a probability a_0 of having zero sons, a probability a_1 of having exactly one, and so on. From these probabilities, you can calculate each male's *expected* number of sons (Section 2.2). If this mathematical expectation is 1 or less, the probability that the name will go extinct approaches one as the number of generations increases. If the expected number is greater than 1, the name can persist forever, but it need not.

The Galton–Watson model was invented to apply to cultural evolution (Section 3.10); it applies to the extinction of family *names*, not to the extinction of *families*, and the passage of a name from father to son is (of course) not due to genetic transmission. Even so, the model *does* apply to changes in gene frequency. If a population is haploid and contains N individuals who have identical fitnesses, and each individual has a different allele at a given genetic locus, then the probability that each allele has of eventually reaching fixation (i.e., 100 percent representation in the population) is $\frac{1}{N}$.[7] The case of sexually reproducing species can be understood in the same way; if there are N individuals in the population and each has a unique pair of alleles at a given locus and the individuals have identical fitnesses, then each allele has a probability of $\frac{1}{2N}$ of going to fixation. The term for zero variation in fitness is "strict neutrality."

It doesn't matter in these two models whether the individuals in the population differ genetically. They could all be clones of each other and it would still be true that the probability that each clone has of being the ancestor of all the individuals in some future generation is $\frac{1}{N}$ (for haploid) and $\frac{1}{2N}$ (for diploid). These two probabilities concern the probability that a given *token* allele (Section 2.5) that exists now will be the ancestor of all the token alleles that exist sometime in the future. The model also applies when allelic types are exemplified by more than one token individual. For example, suppose the A allele has a frequency of 60 percent in a diploid population and the B allele has a frequency of 40 percent, and the two alleles are equal in fitness. This means that A has a 0.6 probability of going to fixation, while B's probability is 0.4. This difference in fixation probabilities is

[7] This simple model assumes that there is no mutational input of new alleles.

compatible with the fact that each of the 2N token alleles at the start of the process has the same probability, $\frac{1}{2N}$, of going to 100 percent.

The Galton–Watson model is about what will happen in the indefinite future, but biologists early in the twentieth century saw a way to conceive of drift as a generation-by-generation process. In the Wright–Fisher model of drift, named for Sewall Wright (1931) and R. A. Fisher (1922, 1930), there are N diploid individuals in the parental generation and the parents each produce the same very large number of gametes from which a sample of 2N is drawn at random to create the next generation of N offspring. Notice that N describes the size of the population in each generation but it also describes the sampling process whereby the next generation is constructed from the one before (Hartl and Clark 2007).[8] If the parental generation is 60 percent A and 40 percent B, the offspring generation has a higher probability of differing appreciably from the parent generation the smaller N is.

5.2 Strict Neutrality

The Galton–Watson model and the Wright–Fisher model were old hat when Motoo Kimura (1968) and Jack Lester King and Thomas H. Jukes (1969) formulated the neutral model of genetic evolution. I take the *strictly* neutral theory to be the thesis that the vast majority of evolutionary changes at the genetic level involve alleles that are *strictly* equal in fitness, whereas the *nearly* neutral theory says that the vast majority involve alleles that are *nearly* equal in fitness. Here I am departing from some of Kimura's verbal formulations in which the distinction between the two theories is a bit blurred. King and Jukes emphasized the radicalness of the theory by calling their paper "Non-Darwinian Evolution." Before the neutral theory was developed, biologists often thought that almost all the genes found in a

[8] Although Wright and Fisher agreed about the mathematics of strictly neutral evolution, they disagreed about the importance of drift in the evolutionary process. In Wright's (1937) shifting-balance model, there are a number of fairly small populations in a species that are semi-isolated from each other; selection moves each of them to their local adaptive peak (Section 4.6), after which drift causes some of those populations to cross an adaptive valley and reach a higher peak. Those populations then send migrants to other local populations, thus improving the fitness of the whole species. Fisher, on the other hand, thought that evolution occurs mainly in very large populations, and so the Wrightian process will have little or no effect.

population evolved because they provided organisms with advantageous phenotypes. The neutral theory was an eye-opener.

Is the strictly neutral theory incompatible with Darwin's idea that selection is the main but not the exclusive cause of evolution? In fact, they are compatible if you restrict Darwin's theory to phenotypes and the neutral theory to genes. It is perfectly possible for phenotypes to differ in fitness though the genes underlying those phenotypes mostly do not. As a simple example, consider a haploid population in which every organism has phenotype A or phenotype B, and these two phenotypes differ substantially in fitness. Suppose there are 1,000 genes in the population that each suffice for an organism to have A and another 1,000 genes that each suffice for an organism to have B. Now suppose that selection drives the A phenotype to 100 percent, with many of the 1,000 genes that code for that phenotype coming along for the ride. Since those alleles don't differ in fitness, they change frequency by a strict drift process. In this scenario, phenotypic selection creates a setting in which random genetic drift takes place.[9]

Testing the neutral theory requires techniques for discerning genetic differences within and among species. When the discipline of population genetics began in the early twentieth century, genetic differences were inferred by finding differences in an organism's phenotype, like eye color or bristle number. The project took a big leap forward in the 1960s when gel electrophoresis allowed biologists to discern differences in the proteins produced by genes (Hubby and Lewontin 1966; Lewontin and Hubby 1966). An even bigger step forward began in the 1980s with the dawn of molecular biology, where new technologies allowed sequencies of nucleotides to be determined; Kreitman's (1983) discovery of variability at the DNA level was an early milestone in that burgeoning enterprise. At each step, scientists were amazed by how much genetic variation there is in populations (Veuille 2019).[10] The strictly neutral theory started out as the hypothesis that all the alleles at a locus have the same fitness, but it morphed into the theory that all the nucleotides at a site are equal in fitness. The latter is the neutral theory of *molecular* evolution.

[9] In similar fashion, True and Haag (2001) use the term "developmental systems drift" to describe how there can be drift at the level of developmental processes, though natural selection governs the evolution of the phenotypic outcomes of those processes.

[10] For useful background on how the concept of gene has changed historically, see Meunier (2022).

Kimura (1968) identified a simple, striking consequence of the hypothesis of strict neutrality. In a diploid population containing N individuals, there are 2N token alleles at a given locus in a generation. If each has a probability μ (this Greek letter is pronounced "mew") of mutating in a generation, and a mutated nucleotide has a probability u of evolving from mutation frequency to fixation, then the expected rate of substitution k (i.e., the expected rate at which mutations originate and then evolve to fixation) at that site is

$$\text{(ERS)} \qquad k = 2N\mu u$$

This expression applies to *all* mutations, regardless of whether they are advantageous, neutral, or deleterious. However, if the 2N nucleotides that exist at a site are equal in fitness, the probability that each has of reaching fixation is

$$u = \frac{1}{2N}.$$

Substituting this value for u into the ERS equation yields the following:

$$(\text{SUB}_{\text{neut}}) \qquad k = \mu$$

The substitution rate is identical with the rate of mutation when there is strict neutrality; the population size N is irrelevant.

What happens if mutations, once they occur, evolve because they are advantageous? Kimura and Tomoko Ohta (1971, p. 467) say that if substitution is carried out by selection for the new mutation, then

$$(\text{SUB}_{\text{adv}}) \qquad k = 4N_e s\mu,$$

where N_e is the effective population size, s is the selective coefficient, and μ is the rate at which advantageous mutants arise. The effective population size is a theoretical quantity whose value is determined by the census size N and other properties of the population, such as sex ratio and degree of inbreeding. The mathematics entails that $N \geq N_e$. In real populations, the inequality is strict. Until the last section of this chapter, it will do no harm to assume that they are the same. The selection coefficient for a mutation M is the fitness of M minus the fitness of W, where W is the wild type allele (i.e., the allele that is prevalent in the population); s is negative if M is deleterious, s is zero if M is strictly neutral, and s is positive if M provides an advantage. The above equation for SUB_{adv} applies only when $s > 0$. Notice that SUB_{neut} is simpler than SUB_{adv}, a point to which I'll return.

5.3 The Molecular Clock

In the early days of the strict neutrality theory, the observation that overall rates of evolution are about equal across contemporaneous lineages was often cited as strong evidence favoring strict neutralism over the hypothesis that natural selection was at work. This equality of rates among simultaneous lineages is called "the clock hypothesis" (Zuckerkandl and Pauling 1965). To evaluate this idea, I first need to clarify why scientists believed that the clock hypothesis is true. Given that scientists don't have time machines, how did they proceed? Consider Figure 5.1. You can infer whether the rates of evolution are the same in the lineages that trace back from species A and B to their most recent common ancestor by observing genetic features of species A, B, and C. You do this by seeing if the sum of the gene frequency differences at multiple loci between A and C (the "genetic distance" between A and C) is the same as the sum of the differences between B and C. Comparing these distances among *current* species tells you about the *past*. The standard inference procedure is this: if the two distances differ *significantly* (a concept I'll discuss shortly), you reject the clock hypothesis; if they do not, you don't reject it. This is the *relative rates test* of the clock hypothesis.

The relative rates test is an example of a general point I made in Chapter 1. In the theory of evolution, different parts of the theory *work together*. The fact of common ancestry means that the *present* features of species A, B, and C provide a portal through which you can ascertain the processes that occurred in the *past*.

Although it is possible to assemble data from current species to test the clock hypothesis, there is a flaw in the idea that strict neutrality entails that rates across simultaneous lineages must be equal. Strict neutrality says that *each* lineage's evolution is determined by *its* overall mutation rate. Neutrality does not entail that the lineages in Figure 5.1 leading from A and B back to their most recent common ancestor had identical mutation rates. Strict neutrality does not entail that the clock hypothesis is true. Kimura and Tomoko Ohta (1971, p. 467) recognize this point when they say "the uniformity of the rate of mutant substitution per year for a given protein may be explained by assuming constancy of neutral mutation rate per year *over diverse lines* (italics mine)." This is a substantive assumption, not a matter of course.

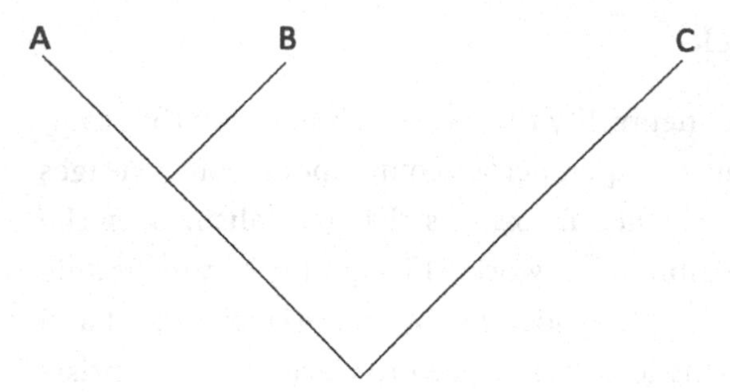

Figure 5.1 The relative rates test.

Kimura and Ohta then introduce the selectionist model of substitution, SUB_{adv}, described above. They think it provides a very implausible explanation of the molecular clock, since it obliges us to

> assume that in the course of evolution three parameters N_e, s, and u are adjusted in such a way that their product remains constant per year over diverse lines. The mere assumption of constancy in the "internal environment" is, however, far from being satisfactory to explain such uniformity of evolutionary rate. In our example of carp-human divergence, we must assume that N_esu is kept constant in two lines which have been separate for some million years in spite of the fact that the evolutionary rates at the phenotypic level (likely to be governed by natural selection) are so different.

Neutralism entails the molecular clock (given the assumption that mutation rates in simultaneous lineages are the same), but selectionism also entails the molecular clock (given the assumption that the product of three parameters are the same in simultaneous lineages). The cases differ, according to Kimura and Ohta, because the assumptions needed for strict neutrality to predict the clock are more plausible than the assumptions needed for selection to do so.[11] The authors conclude that the clock hypothesis is evidence for the neutral theory (Dietrich 1994).[12]

[11] **Question:** Is Kimura and Ohta's argument an application of the Law of Likelihood?

[12] Kimura and Ohta's argument sounds a lot like arguments concerning Copernican and Ptolemaic astronomy before Galileo's telescope discoveries. For discussion, see Sober (2015, pp. 12–22).

5.4 Null Hypotheses

Although the verbal formulation of the strictly neutral theory says that the *vast majority* of genes are strictly neutral, the mathematics of the theory says something more radical. It says that *all* the token alleles at *all* loci have the same probability of going to fixation, $\frac{1}{2N}$, in every diploid population of census size N. To understand this theory, I think the mathematics takes precedence over the informal gloss.

The strictly neutral theory is a "null hypothesis" in the literal sense that it postulates a difference of *zero*.[13] Why do scientists think they need data to test this theory? It seems *highly improbable* that *all* the alleles at *all* loci in *each* present and past population of size N are exactly equal in fitness, especially in view of the fact that it is *highly probable* that at least one phenotype in the entire history of life on earth evolved under the influence of natural selection. To explain why these probability judgments had and continue to have little impact in the neutralism/selection debate, I need to describe some ideas about probability and statistics.

According to Bayesianism, the fundamental goal of inference is to assign probabilities to hypotheses. I mention this because twentieth-century mathematical evolutionary biology developed in the context of an anti-Bayesian statistical philosophy whose first commandment is: *thou shalt not assign probabilities to hypotheses!* The traditional name for this philosophy is *frequentism.* Frequentists hold that it makes no sense to talk about the probabilities of "big" theories like Darwin's theory of evolution, Newton's theory of gravitation, Einstein's theories of relativity, and the neutral theory of genetic evolution.[14] According to frequentists, what does make sense is talking about the probabilities that such theories confer on observations. The distinction

[13] The null-hypothesis concept has been defined in a variety of ways. For example, it sometimes is defined as the hypothesis in a pair of competitors that would be "unremarkable" if it were true. That is *not* the definition I am using.

[14] I use the word "big" here to leave room for the possibility that there are "small" theories that are legitimately assigned probabilities. Consider the theory that says that individual S, who was randomly sampled from the students attending University of Wisconsin this year, has Covid-19. Suppose that S takes a Covid-19 test and it comes out negative. If you know how reliable the test procedure is, and also know the frequency of Covid-19 in the university, you can compute the posterior probability of S's having the disease, given her negative test outcome, by using Bayes' theorem. Frequentists should be okay with this reasoning; what is important to frequentism is the idea that

between Pr(T | O) and Pr(O | T), where O is an observation and T is a theory, is key. As explained in the previous chapter, Pr(T | O) is T's *posterior probability*; that's the probability T would have if O were true. The word "posterior" encourages you to consider the probability that you would assign to T *after* you observe that O is true. Pr(O | T) is T's *likelihood*, meaning the probability that T confers on observation O. I won't discuss broad issues in the Bayesianism/frequentism debate here (but see Sober 2008). Rather, I want to discuss what normative consequences, if any, follow from the fact that a hypothesis is a null hypothesis in the literal sense just described.

Scientists often think that null hypotheses are innocent until proven guilty whereas alternatives to a null hypothesis are guilty until proven innocent. Put differently, the burden of proof is on those who believe the null hypothesis is false; they are the ones who need to provide evidence for their belief. In the absence of such evidence, you can and should believe that the null hypothesis is true. In other words, the null hypothesis should be your *default assumption*.

The idea that null hypotheses are default assumptions played an important role in the controversy about units of selection discussed in Chapter 3. At the start of his influential book *Adaptation and Natural Selection*, George Williams (1966, pp. 4–5) says that adaptation is an "onerous concept," which you should embrace only if the evidence forces you to do so. Williams goes on to explain that he has a three-level hierarchy in mind:

(1) strict neutrality
(2) individual selection and no group selection
(3) individual selection and group selection

When you consider why trait T evolved in a lineage, Williams says that you should start with the assumption that (1) is correct. Only if your data tell you that (1) is wrong should you proceed to (2). And only if both (1) and (2) are ruled out by your data should you proceed to (3). Williams calls this his "ground rule" and "doctrine."[15] He says (p. 18) that this ordering of hypotheses is based on the principle of parsimony.

this unobjectionable calculation of a posterior probability cannot be carried out when the hypotheses you are considering are *not* possible outcomes of a chance process.

[15] **Question**: Does Lewontin and Dunn's (1963) reasoning about the t-allele (discussed in Section 3.7) conform to Williams's protocol?

Here's another context in which it can seem natural to view a null hypothesis as innocent until proven guilty. When you test a drug for its efficacy in preventing a disease in a double-blind randomized controlled trial, the null hypothesis says that

Pr(S has disease D at t_2 | S takes the drug at time t_1) – Pr(S has disease D at time t_2 | S takes a placebo at time t_1) = 0

There may be ethical reasons to think that this null hypothesis should be one's default assumption, but the question of whether null hypotheses should be default assumptions also arises in contexts where ethical considerations don't help. For example, suppose you know that the results of tossing the coin on the table before you are probabilistically independent of each other, meaning that the probability of a coin's landing heads at a given time is the same regardless of whether it landed heads or tails in previous tosses. What you want to discover is whether the coin is fair, meaning that the probability of heads on each toss is ½. That hypothesis can be restated as a null hypothesis since it says that:

(NULL) Pr(the coin lands heads | you toss the coin) – Pr(the coin lands tails | you toss the coin) = 0

If you toss the coin twenty times, there are twenty-one possible outcomes and each has its own probability according to NULL. You can draw a graph with those twenty-one possible outcomes on the x-axis and the probability assigned by the NULL hypothesis to each possible outcome on the y-axis. NULL says that the most probable outcome is ten heads and ten tails, with outcomes becoming increasingly improbable the more they deviate from that even split. If you connect those twenty-one dots, the resulting curve will be bell-shaped. To do a significance test of NULL, you need to use a rule that says when you should reject the hypothesis.[16] The guiding idea is that you should reject NULL if the hypothesis says that the outcome of your tossing experiment was "sufficiently improbable," but how improbable is improbable enough? It is universally agreed that there is no uniquely correct answer to this question.[17] Rather, you need to adopt an arbitrary

[16] Fisher (1925) invented the test of statistical significance in his *Statistical Methods for Research Workers*.

[17] For other criticisms of significance tests, including the fact that they violate the *principle of total evidence*, see Sober (2008, pp. 48–58).

convention. One common choice is to set your *level of significance* at 0.05. This means that you'll reject the null hypothesis if and only if you obtain fewer than six or more than fourteen heads in your twenty tosses. The standard reason for singling out this disjunctive region is that the probability of your twenty-toss experiment yielding a result that falls in that region is just under 0.05 if NULL is true. It is important to see that significance tests aren't Bayesian. You're not figuring out what the probability is of NULL. Rather, you're considering what probability your observations would have if NULL were true.

Notice that you can't do a significance test on the *negation* of the NULL hypothesis:

(DIFF) Pr(the coin lands heads | you toss the coin) – Pr(the coin lands tails | you toss the coin) ≠ 0

What are the probabilities of the possible results in your twenty-toss experiment if DIFF is true? There are infinitely many ways in which DIFF could be true, each assigning its own non-zero probability to what you observe when you toss the coin twenty times. NULL makes definite predictions about the twenty-toss experiment, but DIFF does not. In a significance test, DIFF merely "stands in the wings." It isn't asked to make a prediction of its own; rather, it gets "tested" by virtue of the null hypothesis's being tested.

In view of the inevitable arbitrariness involved in choosing a level of significance, it would be nice if there were a better non-Bayesian approach. There is. The negation of the null hypothesis is DIFF, which can be expressed (more long-windedly) as follows:

(DIFF*) Pr(the coin lands heads | you toss the coin) – Pr(the coin lands tails | you toss the coin) = $\alpha \neq 0$

Here α is an adjustable parameter. Suppose you obtain twelve heads in your twenty-toss experiment. You can use that observation to estimate the value of α. The maximum likelihood estimate is that $\alpha = 12/20 - 8/20$,[18] so the specific instance of DIFF that has the highest likelihood is

(DIFF**) Pr(heads | you toss the coin) – Pr(tails | you toss the coin) = 0.2.

[18] The maximum likelihood estimate of a parameter is the one that makes the observations more probable than any other estimate would do.

DIFF** is the result of fitting DIFF* to the data; DIFF** contains no adjustable parameters, and it makes a definite prediction about the coin tossing experiment just as NULL does.

Frequentist statisticians often evaluate NULL and DIFF* by evaluating NULL and DIFF**. For example, in the likelihood ratio test, you ask whether the difference in probability postulated by DIFF** fits the data *significantly* better than the zero difference in probability postulated by NULL. As with significance tests, the choice of a level of significance in this test is conventional.

The Akaike information criterion (AIC) does not require you to make the arbitrary decision just noted. Hirotugu Akaike (1973) proposed the following measure for evaluating models:

$$\text{AIC}(\text{model M} \mid \text{data}) = \log[\Pr(\text{data} \mid f(M))] - k.$$

The AIC score of model M, given the data, depends on the log-likelihood of f(M), which is the result of fitting model M to data by replacing the adjustable parameters in M with their maximum likelihood estimates, and on k, which is the number of adjustable parameters in the model. The quantity log[Pr(data | f(M))] measures how well the fitted model fits the data. The minus sign in the above equation means that AIC penalizes models for their complexity, where complexity is measured by counting the model's adjustable parameters. Akaike (1973) proved that AIC is an unbiased estimator of how predictively accurate a model will be when its adjustable parameters are estimated from old data and the fitted model is then used to predict new data drawn from the same underlying reality.[19] In the comparison of NULL and DIFF, NULL can't fit the data better than DIFF, but NULL has fewer adjustable parameters than DIFF. AIC says that fit-to-data and parsimony are both relevant to estimating a model's predictive accuracy. It is a striking philosophical fact that parsimony is not an aesthetic frill in AIC; on the contrary, it is epistemically relevant to estimating predictive accuracy (Forster and Sober 1994).[20]

[19] In statistics, an unbiased estimator is one that is "centered" on the true value. Roughly speaking, your kitchen scale is unbiased if and only if repeated measurements of an apple's weight will have the apple's true weight as its average value as you weigh the apple again and again. **Question:** How can the idea of mathematical expectation be used to define what an unbiased estimator is?

[20] The penalty term for complexity in AIC is exactly k; if it were some other number, the resulting estimator of predictive accuracy would be biased.

AIC is an important idea in the area of statistics called "model selection theory," but a better name for that area would be "model *comparison* theory." Unlike significance tests and likelihood ratio tests, AIC provides no criterion for accepting or rejecting a model. Instead, AIC provides *quantitative comparisons*. Null models in the literal sense of that term do have a special status in the AIC framework (in that they are typically more parsimonious than their competitors), but that does not mean that they are innocent until proven guilty.

If default reasoning is to be avoided, what's the alternative? One step forward is to recognize that testing hypotheses is a *contrastive* activity. Instead of setting out to test theory T, you should test T against one or more competitors.[21] Instead of imagining an ordering of hypotheses in which you accept hypothesis H_1 until evidence forces you to reject it in favor of H_2, and you stick with H_2 until the evidence forces you to move to H_3 (as in Williams's protocol for the units of selection problem), a better procedure would be to test H_1, H_2, and H_3 against each other. Burden of proof is a juridical concept; the idea that people charged with crimes should be considered innocent until proven guilty makes sense because there are *ethical* reasons to protect the accused. The situation is different in science if acceptance and rejection are replaced by comparative judgments of *better* and *worse*.

AIC has an interesting bearing on the strictly neutral theory of molecular evolution. You saw earlier, in Kimura's derivation of the substitution rate, that population size is not relevant when there is strict neutrality (i.e., when the selection coefficient $s = 0$). In contrast, a formula for the substitution rate when the mutation is fitter than the wild type needs to have one adjustable parameter for the effective population size and another for the selection coefficient. The strictly neutral model is simpler in the sense that it contains fewer adjustable parameters. According to AIC, the strictly neutral theory *does* have a special epistemic status if it has fewer

[21] There is a special circumstance in which contrastivism about testing is mistaken. If H deductively entails observation statement O, and O turns out to be false, you can conclude immediately that H is false; there is no need to consider hypotheses alternative to H. Likewise, if O entails H, and O turns out to be true, you know that H is true without needing to consider alternatives to H. Contrastivism comes into its own when hypotheses confer non-extreme probabilities on observations and observations don't deductively entail hypotheses.

adjustable parameters than the alternative theories you're considering, but that does not mean that you should accept it until the evidence forces you to reject it.[22] The strictly neutral theory needs to be tested against other models. No model is innocent until proven guilty.

Geneticists often think that the adequacy of the strictly neutral theory should be judged by asking whether you need to invoke natural selection to explain the molecular variation you observe (Veuille 2019). A scientific realist might reply that you *do* need to mention selection if you want your theory to be true. Instrumentalists take a different view of the goal of scientific inference; they say that the goal is to find theories that are predictively accurate. Instrumentalists will not fault the strictly neutral theory for being false if it is more predictively accurate than its competitors. AIC helps underwrite this idea, since AIC says that a model known to be false can be more predictively accurate than a model known to be true. Null hypotheses are often known to be false, but that doesn't mean that they should be tossed in the trash can.[23]

5.5 Alternatives to the Strictly Neutral Theory

The mathematics of the strictly neutral theory not only fails to mention the occurrence of advantageous mutations; in addition, there is no mention of the existence of deleterious mutations. Ohta (1973) proposed a *nearly* neutral theory, which says that almost all of the genetic variation is strictly

[22] Akaike and Kimura were contemporaries and both were senior eminences in Japanese science, but they seem not to have met, nor did Kimura use AIC in his publications. AIC entered evolutionary biology later, in the context of phylogenetic inference, which I'll discuss in Chapter 7. One of Kimura's students, Masami Hasegawa, and his collaborator Hirohisa Kishino, saw how to bring AIC to bear. Kimura was the handling editor of their article, Kishino and Hasegawa (1989). I learned of this history by writing to Masahiro Massuo, who contacted Yusaku Ohkubo, who contacted Masami Hasegawa and Naoyuki Takahata. My thanks to all four.

[23] The relationship of AIC to realism and instrumentalism changes if you define realism to mean that the goal of science is to find theories that are *close* to the truth and instrumentalism to mean that the goal of science is to find theories that are very predictively accurate. It turns out that AIC provides unbiased estimates of distance to the truth if you use the Kullback–Leibler distance measure. That suggests that AIC has realist credentials. However, the fact remains that a model known to be false can have a better AIC score than a model known to be true, which seems to put AIC on the side of instrumentalism.

neutral *or slightly deleterious*.[24] Ohta noted that Kimura's demonstration that strict neutrality entails that the substitution rate equals the mutation rate involves thinking about substitution rates in terms of the number of substitutions *per generation*. However, a generation of large animals is typically much longer than a generation of small animals; for example, the average life span of hominids (humans, chimpanzees, bonobos, gorillas, and orangutans) is between thirty-five and seventy-nine years, whereas the average life span of fruit flies is between six and fourteen days. The strictly neutral theory therefore predicts that fruit flies should have much higher substitution rates (computed per year) than hominids. Biologists in the 1960s and 1970s found that rates of protein evolution are independent of the generation time. Ohta addressed this discrepancy by noting that hominids live in much smaller populations than fruit flies, which means that slightly deleterious mutations increase in frequency and are maintained in human populations more often than they are in populations of fruit flies. Whereas the strictly neutral theory says that population size has no effect on the substitution rate, the nearly neutral theory says that it does. The strictly neutral theory is simpler, but many population geneticists now think that it was too simple, given current observations.[25]

Another challenge to the strictly neutral theory came from biologists who adduced evidence for positive selection on genes. One line of argument focused on *selective sweeps* (Maynard Smith and Haigh 1974). A selective sweep occurs when a new mutation is strongly advantageous and therefore quickly evolves to fixation, with genes on the same chromosome that are spatially close to the favored gene coming along for the ride; this is "genetic hitchhiking."[26] Selective sweeps can be recognized after the fact; a sweep should result in a pair of homologous chromosomes that are unusually homozygous in the region surrounding the advantageous gene. Weaker selection for an advantageous gene takes longer to drive

[24] See Dietrich and Millstein (2008) for a philosophical and historical account of Ohta's work.

[25] Notice that the strictly neutral theory *logically entails* Ohta's nearly neutral theory, but not conversely. The axioms of probability thus ensure that Ohta's theory must be more probable than the strictly neutral theory, regardless of what one observes, if there is a nonzero chance that at least one mutation is slightly deleterious. Population geneticists do not defend Ohta's theory via this Bayesian argument.

[26] The distinction between *selection of* and *selection for* (Section 2.7) applies here.

the gene to fixation, and recombination by crossing over is apt to result in less homozygosity in the adjacent region. A human example of a selective sweep is the evolution of lactose prolongation. Human babies can digest lactose, but the ability disappears in adulthood, or at least that was the situation until the dawn of dairy farming some 10,000 years ago, after which populations in East Africa and in Northern Europe independently evolved genes that make it possible for adults to digest lactose. The different populations evolved different genes that have this effect (Bersaglieri et al. 2004).

The strictly neutral theory is a monolithic and global theory of molecular evolution. Even if strict neutrality turns out to be false in its most universal form, it may still be true for substantial portions of the genomes of different species, but more complex models are needed to take account of the full picture.

5.6 When Drift and Selection Both Affect a Trait's Evolution

Thus far, my discussion of drift may make it seem that selection and drift can't both influence the evolution of the same trait at the same time. After all, the alleles at a locus at a given time either differ in fitness or they do not, and the same point holds for alternative phenotypes. However, biologists now generally agree that drift influences a trait's evolution in smaller populations more than it does in larger, and selection exerts a larger influence when there is more variation in fitness rather than less. More precisely, for a given amount of fitness difference, the influence of drift increases as population size declines; for a given population size, the influence of selection increases as fitness variation increases. The implication is that selection and drift *simultaneously* affect the evolution of a trait.

Here's a two-part analogy. (1) Suppose coin A has a probability of landing heads of 0.4 and coin B has a probability of landing heads of 0.3, and you're going to toss each coin n times. The probability that you'll get *more* heads from tossing A than you'll get from tossing B increases as n increases. (2) Suppose you'll toss coins C and D fifty times each. The probability of getting more heads from C than you get from D is greater if C's probability of landing heads is 0.9 and D's is 0.2 than would be true if C's probability of heads were 0.6 and D's were 0.3.

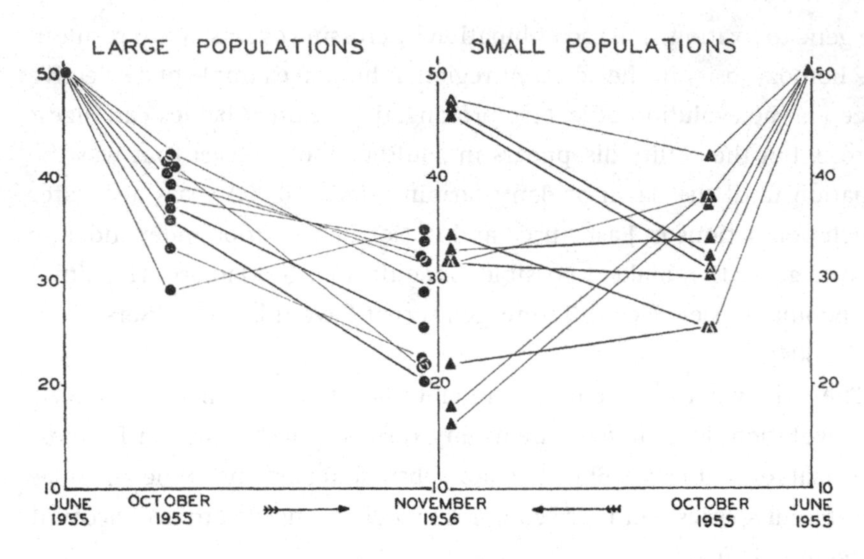

Figure 5.2 An experiment by Dobzhansky and Pavlovsky (1957).

A good example of how selection and drift both influence the evolution of a trait can be found in Dobzhansky and Pavlovsky's (1957) experiment in which a population of fruit flies that had gene A at 100 percent was crossed with a population of flies that had gene B at 100 percent at the same locus.[27] The offspring AB heterozygotes were then used to create 10 small populations (size = 20) and 10 large ones (size = 4,000), each with 50 percent A and 50 percent B. Each new population evolved in its own cage, where the conditions in cages were kept the same as far as possible. The results of the experiment are depicted in Figure 5.2, which comes from Dobzhansky and Pavlovsky's paper. Their caption reads:

> The initial frequency of large replicates is given on the left, and the initial frequency of small replicates, on the right.... The two samples, after 4 generations and after replicates reached equilibrium, are plotted from outside in. The equilibrium frequencies of the large and small replicates are compared on the center axis. (p. 315)

Dobzhansky and Pavlovsky knew that selection would lead gene A to decline in frequency in each population, and that's what they observed. They note that the average reduction of A's frequency in the large populations was

[27] Actually these "genes" were *inversions* but that fine point won't affect the points I'll make.

statistically indistinguishable from A's average reduction in the small. Their main point is that the *variance* in outcomes among the small populations was significantly greater than the variance among the large.

Dobzhansky and Pavlovsky attributed the difference in variance to "the founder effect." When samples are drawn at random from a parent population to found several offspring populations, the offspring populations will differ from each other more if they are small than they would if they were large. The founder effect did not make a difference in the initial frequencies of A and B in the founding populations; rather, it influenced the mix of *other* genes that those populations received, causing the small populations to initially differ from each other more than the large populations did. In Dobzhansky and Pavlovsky's telling, selection explained one set of changes in their experiment (the reduction in frequency of gene A) while the founder effect (a form of drift) explained another (the variation among groups in one treatment compared with the variation in the other).[28]

This experiment shows how drift and selection can both affect the evolution of a single trait, but it was Kimura who developed a mathematical framework for understanding how a gene's probability of fixation is affected by both. Suppose gene W (for wild type) is present at 100 percent in a population, and that gene M then appears by mutation. What is the probability that M will increase in frequency and eventually go to fixation? That depends on the value of the selection coefficient, which can be positive, zero, or negative. With strict neutrality (s = 0), we've already seen that the mutation has a probability of fixation of $\frac{1}{2N}$ in diploids. Kimura (1962, equation 10) proved something more general – that mutation M's fixation probability (u), regardless of whether M is deleterious, neutral, or advantageous, is

$$\text{(FIX)} \quad u = \frac{1-e^{-2s}}{1-e^{-4N_e s}}$$

Here e is Euler's number (≈ 2.71828). Kimura derives this from his more general equation 8, which has three variables – s, N_e, and the allele's frequency p in the population. It says that

$$\text{(FIX*)} \quad u = \frac{1-e^{-4N_e sp}}{1-e^{4N_e s}}.$$

[28] Reisman and Forber (2005) use this experiment to argue that drift is not just an outcome; it is a cause of evolution.

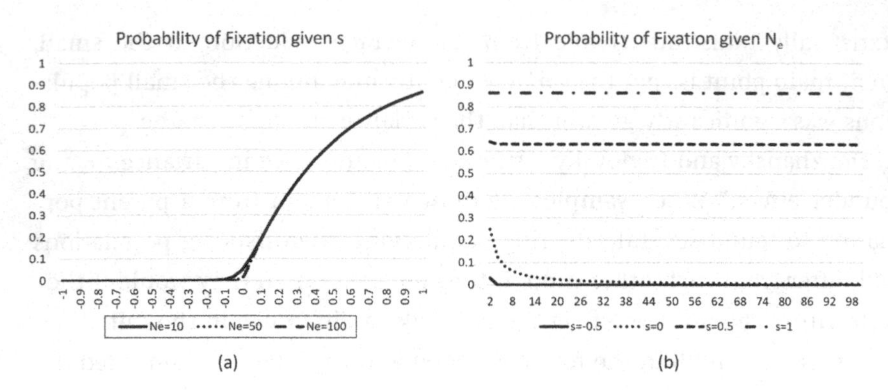

Figure 5.3 The probability of fixation depends on N_e and s.

FIX is the result of setting $p = \frac{1}{2N_e}$ in FIX*. Kimura's derivation of FIX* was complex. In what follows, I'll treat FIX and FIX* as blackbox tools for calculating fixation probabilities.

Figure 5.3 depicts some consequences of FIX.[29] In Figure 5.3a, the probability of fixation increases with s, regardless of the value of N_e.[30] This is not surprising, but matters change when you turn to Figure 5.3b and compare different values of N_e for a single value of s. In this figure, the probability of fixation *decline* as N_e increases, regardless of whether s is positive, negative, or zero. The key to understanding this pattern begins with the fact that Kimura is computing the probability of fixation of a newly minted *token* allele. This means that as you consider larger and larger values of N, the starting frequency of the allele automatically declines. This helps explain why evolving from a frequency of $\frac{1}{2N}$ to 100 percent in a population of 10,000 is less probable than evolving from $\frac{1}{2N}$ to 100 percent in a population of size 10 when the s values are the same. This pattern is also visible in Figure 5.3a if you look at values of s that are slightly greater than zero. The fact that FIX computes the fixation probability of a token allele in a population of size N_e also helps explain why the fixation probability for s = +1 in Figure 5.3b does not asymptotically approach unity as N_e gets larger

[29] My thanks to Matthew J. Maxwell for preparing this figure and helping me understand it.

[30] The lines in Figure 5.3a don't overlap exactly.

and larger. Even when $s = +1$ and N_e is huge, there is a nonzero probability that the token allele introduced by a mutation in one generation fails to be present in the next. This holds no matter how big N_e is.[31]

These points about Figure 5.3 throw light on the rule of thumb I've been using. The rule says that when selection is in control of the evolutionary process, fitter traits increase in frequency and less fit traits decline. In this context, I've said that selection's control over evolutionary outcomes is greater the bigger N is. How can this be, if FIX is right in saying that positive selection coefficients, no matter how large, always fail to ensure that the fitter trait goes to fixation? The answer is that the rule of thumb should not be understood in terms of Kimura's FIX formula. FIX has only two independent variables, s and N_e, and so it doesn't allow you to explore how the fixation probability would change if you held the initial trait frequency p constant and increased the value of N_e. This is important, as the rule of thumb needs to be evaluated by answering these two questions:

- What happens to the fixation probability u if p and s are held fixed and Ne increases?
- What happens to the fixation probability u if p and Ne are held fixed and s increases?

Here p is the allele's frequency at any value between 0 and 1 you want to consider. With respect to the first question, the probability of fixation increases as N_e increases when $s > 0$, and the probability of the allele's disappearing from the population increases as N_e increases when $s < 0$. With respect to the second, for any value of N_e, the probability of fixation goes up as s increases. These claims are consequences of Kimura's FIX*.

5.7 Drift as Cause and Process

It is important to recognize that the terms "drift" and "selection" are each subject to a process/product ambiguity (Millstein 2002a, 2006;

[31] Matters change if the same type of mutation happens again and again. With repeated advantageous token mutations of the same type, the probability that one of them will reach fixation approaches 1 even if each has a small probability of doing so.

Stephens 2004; Reisman and Forber 2005; Shapiro and Sober 2007; Dietrich and Millstein 2008). This is to be expected, as many other terms are in the same boat (Sober 1984, 1991a). "Marriage" can denote the process of getting married, but it also can refer to the product of that process, the state of being married; "erosion" can denote the oozing of mud down a hill, but it also can refer to the mud puddle at the bottom. The process/product distinction, and the fact that the drift process is probabilistic, make it clear that a drift process can occur without the population's trait frequencies changing (contrary to Brandon 2005). True, the expected behavior of a pure drift process is that trait frequency does a random walk, but it is possible (though improbable) that there is no walk at all.

Beatty (1984, 1992) and Millstein (2002a, 2021) define selection as discriminate sampling and drift as indiscriminate sampling. This is a dichotomy. In discussing Kimura's FIX equation, I said that the value of N represents the strength of drift and the value of s represents the strength of selection. Here selection and drift are described quantitatively, not in terms of a dichotomy. The dichotomous view and the quantitative view conflict, since the discriminate/indiscriminate distinction makes no mention of population size.

Millstein (2021, p. 26) contrasts mathematical and causal definitions of drift; she criticizes the definition of drift in terms of population size on the ground that it doesn't treat drift as a causal process. Millstein (p. 21) says that "population size does not seem to be a process." She is right that N is a number, not a process, but the point is that N represents a causal factor that affects evolutionary outcomes. This is clear in the Wright–Fisher model described earlier. In that model, N represents the number of diploid individuals in a generation, but 2N represents the process of random gamete sampling that results in the next generation of N diploid individuals.

Table 5.1 describes a few consequences of the idea that the strengths of selection and drift can be understood by talking about what happens when you vary the value of either s or N while holding fixed the value of the other. If drift and selection are distinct causes, must it be possible for each to occur without the other? I think not. If infinite population size is physically impossible, selection cannot occur without drift. I accept that

Table 5.1 *Selection and drift are causes of evolution. When does each occur?*

	Finite population	Infinite population
Variation in Fitness	Selection and drift	Selection only
No Variation in Fitness	Drift only	Neither

consequence. For drift and selection to be distinct causes, it is enough that s (or N) can vary while N (or s) remains the same.

My talking about infinite population size might sound bizarre. How can it make sense to think about trait frequencies in an infinite population? Evolutionary biologists talk about infinite population and philosophers of biology have followed suit, but that doesn't remove the feeling of uncanniness. In Table 5.1, what you need to think about is just the *limit* of what happens as N is made larger and larger; this limit is defined by what happens in a series of increasingly large but always *finite* populations. There are no infinite populations *in* this series; rather, infinite population size is the *limit of* that series. In similar fashion, someone might say that a fair coin would land heads half the time if it were tossed an infinite number of times. If this sounds too weird, you need to remember that this talk of infinity is shorthand. What is true is that the probability of getting however close you want to 50 percent heads approaches one as the number of tosses is made larger and larger.

Millstein (2021) notes that the thesis that selection and drift involve different types of sampling process allows for the possibility that a single trait's evolution can be influenced by both. She cites the example of Maxime Lamotte's (1959) work on land snails in which he concludes that camouflage provides a selective advantage but also notes that the founding of new populations is important because of "chance variations in the composition of the first colonizers." Notice that selection and drift in this example are said to be *sequential*, not *simultaneous*. The same interpretation applies to the Dobzhansky and Pavlovsky experiment discussed earlier if the initial creation of the small and large populations is a pure drift process and the subsequent evolution of the populations is due solely to selection. The latter claim goes wrong, however, because it ignores the role of population size in the populations' evolution *after* they were formed.

This raises a new question: when drift and selection *simultaneously* affect a trait's evolution, should you conclude that selection and drift are separate processes? Consider this analogy: When you toss a coin some number of times, the probability of getting a given frequency of heads depends on the probability of heads on each toss and also on the number of tosses. There are two causal *factors* at work here, but it seems odd to say that there are two causal *processes*. The more natural interpretation is that there is a single process – tossing the coin repeatedly – that has two causally relevant properties. So it is with selection and drift when they occur simultaneously (Sober 1984, p. 115; Abrams 2007; Gildenhuys 2009).

This conclusion should be grounded in a general criterion that says how causal processes are individuated. I admit that I can't supply that criterion. However, I think that those who think that selection and drift are distinct processes in this case also need a criterion. There is philosophical work to be done here; neither side has the burden of proof.[32]

5.8 Selection, Drift, and Domination[33]

Since selection (represented by the selection coefficient s) and drift (represented by the effective population size N_e) both influence a mutation's probability of reaching fixation, the question naturally arises of whether selection exerted a bigger influence than drift in the evolution of a given allele or nucleotide in a given population. It is clear that $s = 0.4$ involves a stronger selective push than $s = 0.2$, and $N_e = 100$ means that drift exerts a stronger influence than it would if $N_e = 10{,}000$, but what does it means for selection to be *a stronger cause than* drift in the evolution of a given allele or nucleotide? That is, when does selection *dominate* drift?

On its face, this question resembles questions that might be raised about some of the examples discussed earlier:

[32] See Dupré (2012) for discussion of the advantages of a process ontology in biology.

[33] This section reproduces material from Maxwell and Sober (2023b).

- If viability selection favored plain peacock tails over gaudy tails, while sexual selection favored gaudy tails over plain ones, which of these causes exerted the stronger influence on the evolutionary outcome? (Section 2.2)
- If individual selection favored selfishness and group selection favored altruism in a given metapopulation, which of these causes exerted the stronger influence on the evolutionary outcome? (Section 3.4)

In each of these examples, two causes push in opposite directions, and you can infer which was stronger by seeing how the population evolved. If gaudy tails increase in frequency, that is evidence that sexual selection exerted the stronger push. If selfishness increases in frequency, that is evidence that individual selection exerted the stronger push. Is comparing the strengths of selection and drift different? Perhaps it's like the proverbial comparison of apples and oranges.

Biologists have suggested different criteria for comparing the impact of selection and drift. Kimura (1968) proposed that the criterion for drift's dominating selection should be $sN < ½$. Li (1978, p. 374) argued that "a more reasonable definition" would be $sN < 1$, although he notes that focusing on the distinction between $sN < 1$ and $sN > 1$ is "somewhat arbitrary." Roughgarden (1979, pp. 76–78) says that when $4N_es << 1$, "we might say that drift has overpowered selection," and when "$2N_es >> 1$, the influence of selection is conspicuous." Setting aside Roughgarden's "<<," I want to consider criteria that say drift dominates selection precisely when $N_es < c$ (for some positive constant c). I'll call these "N_es criteria."

As stated, N_es criteria have an odd feature; they classify cases in which N_es is very negative as cases in which drift dominates selection. This is no news to population geneticists, who routinely follow Kimura's (1983) formulation, which says that drift dominates selection precisely when the absolute value $|N_es| < c$. With this slight modification, N_es criteria judge that selection dominates drift when N_es is positive and very big and also when N_es is negative and very far from zero. In both cases, selection is said to be in the driver's seat. What may be surprising is that if you use Kimura's FIX formula to calculate probabilities of fixation for different values of $|N_es|$, two alleles that have the same value for $|N_es|$ can differ drastically in their fixation probabilities. This can

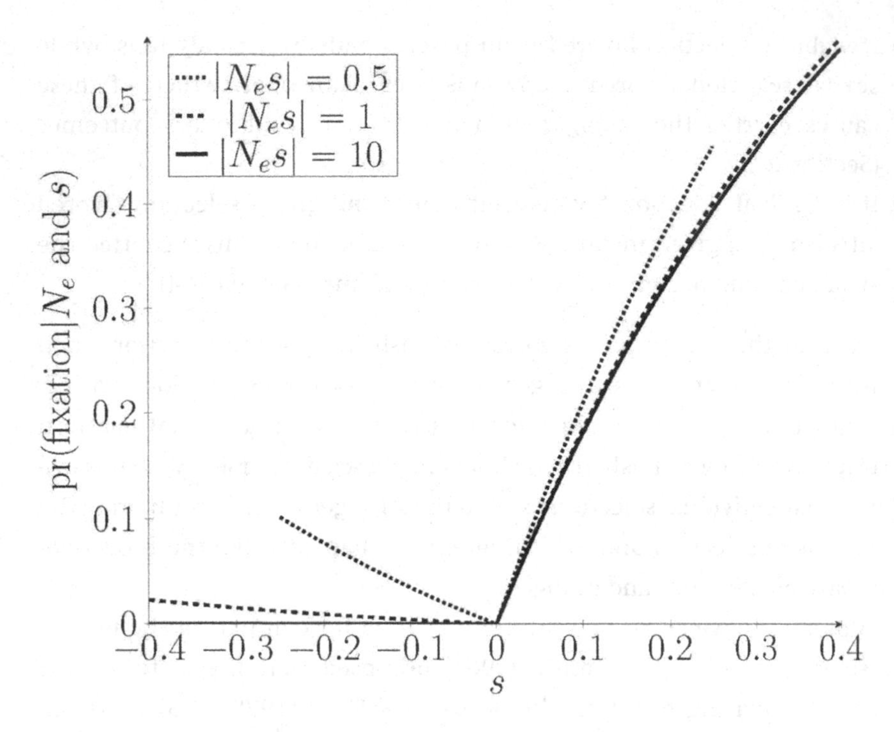

Figure 5.4 Points on the same curve have the same value for $|N_e s|$, but the fixation probabilities associated with those points differ. Distinct curves with the same value for $|N_e s|$ have different fixation probabilities.

happen when the alleles have the same positive $N_e s$ value, when they have the same negative value, and when one allele has a positive value and the other has a negative. Figure 5.4 depicts some examples. This pattern arises because N_e and s together influence the fixation probability in FIX, but the functional form of that determination is not solely by way of the product $N_e s$, since the selection coefficient stands alone in the numerator.

A better criterion for drift's dominating selection is possible. Suppose an allele or nucleotide that evolved to fixation had a probability u of doing so, given the actual positive value of its selection coefficient s and the actual effective population size N_e. A natural way to decide whether drift was a stronger cause than selection would be to compare u with the probabilities of fixation in two counterfactual situations. The first involves holding fixed the actual population size and the actual initial trait frequency, and

Table 5.2 *When does drift dominate positive (or negative) selection? Probabilities of fixation under three conditions*

		Effective Population Size	
		Finite	Infinite
Selection coefficient	Positive (Negative)	u	1 (0)
	Zero	$\frac{1}{2N}$	

thinking about what the fixation probability would have been if there had been no selection; in that circumstance, the probability of fixation is $\frac{1}{2N}$. This comparison is standard among population geneticists. The second, but equally important, comparison involves holding fixed the actual value of the positive selection coefficient and the actual initial trait frequency, and thinking about what the fixation probability would have been if there had been no drift; as N gets larger and larger, the probability of fixation asymptotically approaches 1 if s is positive.[34] Values for these two counterfactual probabilities and the actual probability u are represented in Table 5.2. As before, talk of what happens in an infinite population is shorthand for what happens in the limit as (finite) N values increase. The ordering of these probabilities is $\frac{1}{2N} < u < 1$. Positive selection raised the probability of fixation; drift diminished it.

The three probabilities just described – one of them actual and two of them counterfactual – allow us to say that drift was a stronger cause of the fixation event than positive selection precisely when

$$1 - u > u - \frac{1}{2N},$$

which is equivalent to

$$\text{(DDPS)} \quad u < \frac{\frac{1}{2N} + 1}{2}.$$

This criterion says that drift dominates positive selection precisely when the actual probability of fixation is less than the average of the

[34] As explained in Section 5.6, the value of unity is not obtained by using Kimura's FIX formula. Here I am using FIX*.

two counterfactual probabilities. Put differently, drift dominates positive selection precisely when u is closer to the fixation probability under "pure drift" than it is to the fixation probability under "pure positive selection."

A second analysis is represented in Table 5.2, one that compares drift with negative selection. The ordering of the entries in that case is $\frac{1}{2N} > u > 0$, and the criterion for drift's dominating negative selection is that

$$\text{(DDNS)} \quad u > \frac{1}{4N}.$$

Notice that DDPS and DDNS have census size, not effective population size, on the right-hand side.

The two criteria for drift's dominating selection are an application of a more general idea – that you can determine which of two causes contributed more to a given effect by *zeroing-out* each while holding the other fixed. This terminology may be somewhat new (it's from Clatterbuck et al. 2013), but the idea is old, in both science and philosophy. Precedents in science can be found in the use of norms of reaction to describe how genes and environment affect phenotypes (Lewontin 1974). Zeroing-out is also part of the rationale for randomized controlled trials In philosophy, theories of probabilistic causality use the zeroing-out idea (Eells 1991), and Sober (1988a) and Wright et al. (1993) discuss a "counterfactual test" for comparing causal efficacies. Doubtless there are earlier precedents in science and philosophy.

Figure 5.5 summarizes DDPS and DDNS by using the FIX formula to compute fixation probabilities for different possible values of s and N. The result is that drift dominates selection in the evolution of a nucleotide or allele if and only if the values of s and N fall in the grey region of the figure.[35] The boxiness of that region may come as a surprise. Although the values of N represented are unrealistically small, the grey region gets only a tiny bit narrower as N increases. In fact, the gray region remains gray for *all* values of s between s = 0 and $s = \frac{-\ln\left(\frac{1}{2}\right)}{2}$ (≈

[35] The criterion for drift's dominating selection proposed here conflicts with the conclusion drawn in Clatterbuck et al. (2013), that the strengths of selection and drift are incommensurable. Those authors consider a zeroing-out analysis but judge it a failure. Brandon and Fleming (2014) criticize Clatterbuck et al. and defend an $N_e s$ criterion.

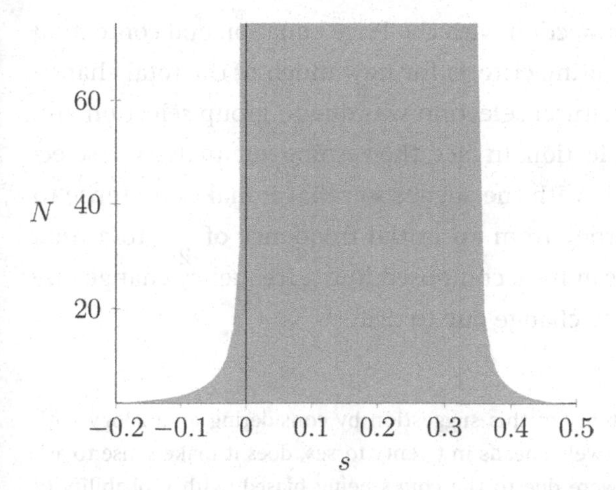

Figure 5.5 According to the method of zeroing-out, drift dominates selection in the evolution of an allele now at fixation precisely when its values for s and N fall in the grey region.

0.35), *regardless of how big N is.*[36] Zeroing-out contrasts dramatically with $N_e s$ criteria. This does not mean that $N_e s$ criteria are useless; rather, the point is that they don't characterize what it is for drift to be a stronger cause than selection.

Since the derivation of the numerical picture shown in Figure 5.5 uses Kimura's FIX equation, the derivation assumes that there is no dominance or epistasis in fitnesses (concepts that I'll explain in Section 8.3) and that effective population size and selection coefficients don't fluctuate during the course of an allele's evolution. The zeroing-out strategy may be useful in thinking about what it means for drift to dominate selection when these and other idealizations are relaxed.[37]

It is interesting that the zeroing-out approach to comparing selection and drift does not provide the sort of quantitative decomposition I described at

[36] Plugging possible values of N into FIX may seem like a mistake, since FIX has N_e as a variable, not N, but Figure 5.5 can be interpreted by taking that fact into account. Since $N_e < N$, you can correct for the discrepancy by moving from a higher value of N on the y-axis to one that is lower. The boxiness means that that downshift makes little difference.

[37] Maxwell and Sober (2023b) discuss zeroing-out further and also discuss Rice's (2004) and Cutter's (2019) justifications of an $N_e s$ criterion. I am not arguing that such criteria are useless, only that they do not do a good job of comparing the causal power of selection and drift.

the end of Chapter 3 in connection with the Price equation and contextual analysis, which offer competing criteria for how much of the total change in trait frequency due to natural selection was due to group selection and how much to individual selection. In fact, the zeroing-out analysis of selection and drift is compatible with the suggestion that it makes no sense to say that a mutation's journey from an initial frequency of $\frac{1}{2N}$ to a final frequency of 100 percent can be decomposed into a frequency change due to selection and a frequency change due to drift.[38]

[38] **Exercise:** Assess the plausibility of this suggestion by considering an analogy with coin tossing. When you get twelve heads in twenty tosses, does it make sense to ask how many of those heads were due to the coin's being biased (with probability of heads = 0.8) and how many were due to there being twenty tosses?

6 Mutation

Although Mendelian genetics came after Darwin, Darwin advanced a substantive claim about the variants on which the process of natural selection goes to work. He thought that these variants occur "at random" and that evolution by natural selection involves the accumulation of a large number of small phenotypic changes, rather than the accumulation of a small number of large changes. These two Darwinian ideas – the randomness of mutation and the gradualism of adaptive evolution – are the topic of the present chapter.[1]

6.1 Darwin and Asa Gray

In Section 1.8, I quoted the analogy that Darwin drew between beneficial mutations and the stones that fall from a cliff, some of which an architect selects to construct a building. The shapes and sizes of the fallen stones have their causes, but the explanation for why the stones have those features is not that they would later prove useful. Darwin's thesis is that mutations also have their causes, but they do not occur because they would be useful to the organisms in which they occur. This conception of

[1] This formulation of the thesis of gradualism with its double use of "small" and "large" raises the question of what these vague terms mean. I discuss that in Section 6.5. Note also that the gradualism I'm describing is distinct from what Eldredge and Gould (1972) call *phyletic* gradualism, which is the claim that the rate of evolution in a species' lifetime is small and constant. Eldredge and Gould think the usual pattern is rapid evolution in the initial 10 percent of a species' lifespan followed by stasis thereafter; this is why they use the phrase "punctuated equilibrium" to name their theory. They think their picture and Darwin's are different, in that Darwin constantly stressed that evolution is slow. However, there is room in the Darwinian picture for rates varying, both within lineages and between them, as can be seen in Darwin's Figure 1.1 diagram.

Figure 6.1 Darwin's idea that mutations are "random."

randomness is depicted in Figure 6.1. The line represents different possible values of a quantitative phenotype – for example, blood pressure. The dot represents an organism's present phenotypic state. Suppose a mutation will move the organism from its present location to one of the points on the line and that O_1 and O_2 represent two possible phenotypes that might be optimal. To keep things simple, I'll assume that the fitness of a mutation is a linear function of how close it is to the optimum. For any given mutation size, mutating to the right would be better than mutating to the left if O_2 were the optimum, and the reverse would be true if O_1 were the optimum. According to Darwin, the probability a mutation has of yielding a given phenotypic outcome is the same regardless of whether the optimum is O_1 or O_2. There is no assumption here that a mutation to the right will have the same probability as a mutation to the left, nor that the amounts of phenotypic change induced by left and right mutations would be the same.[2] Darwin's idea remains the received view in modern biology (Lenski and Mittler 1993).

Asa Gray, Darwin's long-time friend and the foremost North American proponent of his theory, embraced the hypothesis of natural selection, but he was greatly troubled by the thesis that mutations are random. In the review he published of the *Origin*, Gray (1860) urged Darwin "to assume, in the philosophy of his hypothesis, that variation has been led along certain beneficial lines." Darwin wrote to Gray in response:

> Yesterday I read over with care the third Article; & it seems to me, as before, admirable. But I grieve to say that I cannot honestly go as far as you do about Design. I am conscious that I am in an utterly hopeless muddle. I cannot think that the world, as we see it, is the result of chance; & yet I cannot look at each separate thing as the result of Design. To take a crucial example, you lead me to infer that you believe "that variation has been led along certain beneficial lines." I cannot believe this; & I think you would have to believe, that the tail of the fan-tail [pigeon] was led to vary

[2] For a more wide-ranging discussion of what Darwin meant by "chance," see Gigerenzer et al. (1989, pp. 132–141).

> in the number & direction of its feathers in order to gratify the caprice of a few men. Yet if the fan-tail had been a wild bird & had used its abnormal tail for some special end, as to sail before the wind, unlike other birds, everyone would have said what beautiful & designed adaptation. Again I say I am, & shall ever remain, in a hopeless muddle (Darwin 1860).

The dilemma described here is the apparent impossibility of reconciling the randomness of mutation with the idea that God created the universe. When Darwin twice says that he is in a muddle, is he dissembling out of modesty and politeness? This private expression of confusion did not find its way into the *Origin*. Instead, there one finds a confident endorsement of *deism* – the thesis that God created the universe, the laws of nature, and the initial conditions of the universe, after which God was hands-off. All subsequent events that occur in nature are consequences of what God did at the start. Darwin found this conception of God more attractive than the idea that God repeatedly reaches into nature to get events to occur. He announces his deism at the very start of the *Origin* by quoting a philosopher who was his Cambridge teacher:

> But with regard to the material world, we can at least go so far as this – we can perceive that events are brought about not by insulated interpositions of Divine power, exerted in each particular case, but by the establishment of general laws. — William Whewell, *Bridgewater Treatise*.

Darwin reaffirms this thesis near the end of the book when he says:

> To my mind it accords better with what we know of the laws impressed on matter by the Creator, that the production and extinction of the past and present inhabitants of the world should have been due to secondary causes, like those determining the birth and death of the individual (Darwin 1859, p. 488).

By "secondary causes," Darwin means the causes that came into play after God (the first cause) created the universe.[3]

It wasn't just philosophical theology that led Darwin to deny that mutations arise because they will be beneficial. His detailed study of artificial selection, described in his 1868 book *The Variation of Animals and Plants Under*

[3] Darwin's deism was gradually replaced by agnosticism; the argument from evil was a compelling motivator (Sober 2011b). Clatterbuck (2022) argues that Darwin abandoned deism after the *Origin* appeared because of his conviction that mutations are unguided.

Domestication, was an important influence. Darwin repeatedly observed that variation is abundant. If the pigeons in a population have beaks that have an average length of two inches, there will inevitably be variation around that mean value. If one breeder wants to create a variety that has longer beaks while another wants to create a variety that has shorter beaks, both will succeed. Similarly, a single breeder can create a variety with longer beaks and then reverse course and shorten them. According to Darwin, breeders *exploit* pre-existing variation; they do not *induce* it (Darwin 1868, pp. 419–421).[4]

Darwin's disagreement with Gray about the randomness of mutation set the tone for subsequent discussions of this question. The question of whether mutations are random and the question of whether God intervenes in the evolutionary process were often linked. In Darwin's and Gray's shared formulation, the alternative to *random* mutations is that mutations are *guided*, and "guided" means that mutations are intentionally caused by an intelligent designer. Although the random/nonrandom distinction and the mindless/mindful distinction were linked historically, they need to be disentangled. Indeed, Darwin provides a precedent for doing so. Natural selection is mindless, but it is also nonrandom, in the sense that populations have a greater probability of evolving towards the optimum than they have of evolving away from it when natural selection governs their evolution. We can and should consider whether nonrandom mutations exist without assuming that nonrandom mutations must be guided by an intelligent designer.

6.2 Two Experiments

The processes that induce mutation aren't *far-sighted*. The fact that the environment will get colder in 100 years does not cause bears now to step up their production of mutations that make for thicker fur. The processes that cause mutations are like other causal processes, in that causes precede their effects. However, that does not rule out the possibility that the *current* environment can induce a shift in mutation probabilities "along

[4] In between publishing the *Origin* and *Variation of Plants and Animals*, Darwin spent eight years writing his four-volume treatise on barnacles, the study of which increased his confidence in the ubiquity of small variations (Ruse 1979, Mannouris 2011).

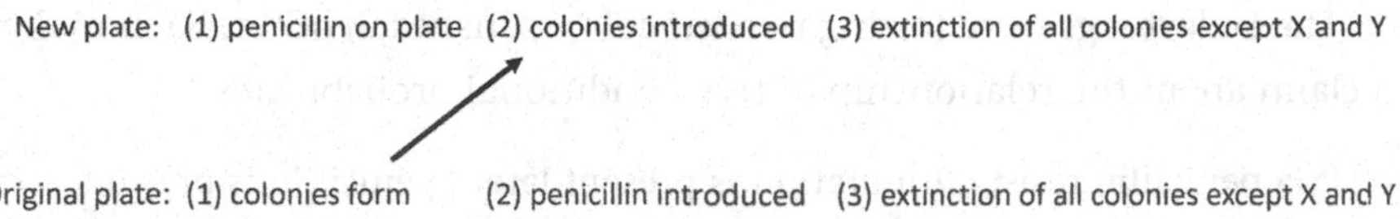

New plate: (1) penicillin on plate (2) colonies introduced (3) extinction of all colonies except X and Y

Original plate: (1) colonies form (2) penicillin introduced (3) extinction of all colonies except X and Y

Figure 6.2 The Lederberg and Lederberg (1952) experiment.

beneficial lines." If the weather *now* gets colder, does that *now* increase the probability of mutations that cause thicker fur?

Esther Lederberg and Joshua Lederberg (1952) ran an experiment to test the hypothesis that mutations are adaptive responses to problems. Just as colder weather is a problem for bears, so penicillin is a problem for E. coli. When E. coli encounter penicillin, does this cause them to produce mutations that help them survive the toxin? The Lederbergs' experimental design is depicted in Figure 6.2. E. coli were added to a plate (call it the "original") on which there is no penicillin. The E. coli then reproduced, forming numerous spatially separated colonies, two of which I'll call "X" and "Y." That plate was then stamped with a cloth, which the Lederbergs then used to stamp a new plate on which penicillin was already present, thereby transferring to the new plate a copy of the E. coli colonies that were present on the original. The result on the new plate was that all the colonies, except for X and Y, went extinct. Was this because the penicillin on the new plate caused the X and Y colonies to produce adaptive mutations, whereas the other colonies on that plate were unable to do so? The Lederbergs argued that the answer is *no*, based on what happened to the original plate. After the original plate was stamped to create the new plate, penicillin was added to the original plate. The result was that all the colonies on the original plate died except for X and Y. The difference between the two plates – whether colonies exist on a plate before or after penicillin was introduced – made no difference in the outcome.[5]

[5] **Question:** The Lederbergs' experiment suggests that the X and Y colonies on both plates had something in common that allowed them to fight back against penicillin and that this commonality was on the original plate before that plate was exposed to penicillin. Does the experiment discriminate between the following two hypotheses?

(H_1) The X and Y colonies on the two plates both had genes that confer penicillin resistance, and these genes were present on both plates before they were exposed to penicillin.

(H_2) X and Y had genes in common that cause new mutations to occur at other loci that confer penicillin resistance once there is penicillin in the environment.

The Lederbergs were testing two models against each other. Each makes a claim about the relationship of two conditional probabilities:

Pr(a penicillin resistant mutation is present later | penicillin is present earlier)
Pr(a penicillin resistant mutation is present later | penicillin is absent earlier)

The random mutation hypothesis is a null hypothesis; it says the difference between the two probabilities is zero. The adaptive mutation hypothesis says that the first probability is bigger than the second.

The Lederbergs considered one environmental problem (penicillin) and one possible adaptive response (mutations that confer penicillin resistance). I now want to describe a hypothetical experiment in which there are two problems and two possible adaptive responses (Sober 2014). Consider a species of blue organisms; in the experiment, you take a large number of them and put them in a red environment and take an equally large number and put them in a green environment. In each of these two settings, the organisms gain a selective advantage from matching their environments. Red organisms out-survive green organisms in a red environment, but the reverse is true in a green environment. This is because you allow organisms that match their environment to survive and reproduce and you prevent non-matching organisms from doing so. In this experiment, you monitor how often the blue organisms mutate to red and how often they mutate to green in each of the two environments.[6] Since mutation probabilities are small, the experiment needs to involve a very large number of blue organisms so that you get mutation frequencies that are sometimes positive. As shown in Table 6.1, the observed outcome of your experiment is four frequencies; they need not sum to 100 percent.

Is there an observation one could make of the plates in the experiment that would discriminate between these two possibilities? The Lederbergs' experiment was inspired by Luria and Delbruck's (1943) fluctuation test (see Bois 2016 for a succinct exposition). Does that paper suggest an answer to the above two questions?

[6] This experimental design relies on scientists being able to identify mutations when they occur. Historically, this has been a highly nontrivial problem. If a blue organism turns green, that may or may not be due to a mutation's occurring. I'm assuming here, for the sake of keeping the example simple, that color changes are caused by mutations.

Table 6.1 *If blue organisms are put in red and green environments, you can observe how frequently they mutate to red and how often they mutate to green in each environment*

	red environment	green environment
red mutation	f_1	f_2
green mutation	f_3	f_4

A model that postulates guided mutations postulates the following two probabilistic inequalities:

Pr(red mutation | red environment) > Pr(red mutation | green environment)
Pr(red mutation | red environment) > Pr(green mutation | red environment)

These two inequalities are logically independent of each other. The first says that blue organisms have a higher probability of mutating to red in a red environment than they have if they are in a green environment. The second says that, in a red environment, red mutations have a higher probability of arising than do green mutations. Both inequalities are needed to fully express what guided mutation means. The first, by itself, could be true just because red environments are more mutagenic than green environments. The second, by itself, could be true just because red mutations are more probable than green ones, regardless of the environment. To complete the formulation of what the guided model says, you need to add two further inequalities:

Pr(green mutation | green environment) > Pr(green mutation | red environment)
Pr(green mutation | green environment) > Pr(red mutation | green environment)

Table 6.2 summarizes the four inequalities that the model of guided mutations proposes.

Given that the hypothesis of guided mutation is defined by the inequalities shown in Table 6.2, there are several competing models, each of which denies one or more of the inequalities. The simplest of these deniers is a null model. It says that all the probabilities in Table 6.2 are equal. Here I'm using the term "null" in the literal sense discussed in Section 5.4.

The null model is simpler than the guided model in that the null model has no adjustable parameters whereas the guided model has four. Each of

Table 6.2 *The guided mutation model asserts four probabilistic inequalities. Each probability has the form Pr(mutation | environment).*

	red environment		green environment
red mutation	p_1	$>$	p_2
	$\vee$		$\wedge$
green mutation	p_3	$<$	p_4

the inequalities postulated by the guided model corresponds to an adjustable parameter; for example, when the guided model says that $p_1 > p_2$, you should think of this as the claim that $p_1 - p_2 = \alpha_{1,2} > 0$, where $\alpha_{1,2}$ is an adjustable parameter. The difference in simplicity between the two models is relevant to assessing which of these two models is more predictively accurate, or so says AIC, the Akaike information criterion (Section 5.4). That doesn't mean that the null model is automatically better. Whether it is better depends on the data.

The kind of experiment I've just described can be carried out in numerous species and in numerous environments, with different experimental manipulations determining which mutations are beneficial. One feature of the experiment is easily relaxed; there are just two environments in the experiment I've described, but the experimental design can and should be changed from two-by-two to n-by-n. Biologists now expect that models that postulate guided mutation will repeatedly be bettered by null models. Nonetheless, let's consider the possibility that there is a species somewhere and a set of environmental manipulations that will yield the reverse result, with the guided model bettering the null. What should biologists say about that result if it occurs?

The right reaction, I suggest, is for biologists to investigate how the organisms manage to produce mutations along beneficial lines. This is a scientific problem, akin to the problem of figuring out how planaria that are cut in two manage to regenerate their missing halves. Part of the problem is to figure out how this adaptive response evolved. Jumping to the conclusion that the mutations were produced by intelligent design would be premature. This reflects my earlier comment that guided mutation does not entail an intelligent designer.

The color-matching experiment provides observational data that allows you to test a guided model against a null model. The observations do the work here, so the line of reasoning I've described might be called "bottom-up." A complementary approach is top-down; there are theoretical reasons why it would be hard for organisms to evolve devices that increase the probability of mutations that happen to be advantageous in one environment, while reducing the probability of those same mutations when they happen to be disadvantageous in another. How is an organism supposed to "tell" which mutations will help and which will hurt in its present environment? Does it survey its contemporaries and record which of them survive and reproduce more successfully, determine which phenotypes have allowed those individuals to do so, and then single out which of its own genes, once mutated, would produce those phenotypes in its progeny, and then cause those mutations to happen? Alternatively, organisms might have, lodged in their genomes, a "memory" of which genes worked well in which circumstances. With this ancestral record, and with detailed information about the present environment, the organism could preferentially produce advantageous mutations.[7] It will come as no shock if organisms lack these abilities; they do so for the same reason that zebras don't have machine guns with which to repel lion attacks (Krebs and Davies 1981). Selection acts only on variants that are actually represented in the population. An additional consideration is that many organisms are *phenotypically plastic*; that is, they are able to modify their phenotypes without changing their genes. When the climate gets colder or warmer, bears grow thicker or thinner fur without new mutations arising that induce this shift. Organisms don't *need* guided mutations when phenotypic plasticity does the job, and the cool thing about phenotypic plasticity is that it induces changes *faster*.

6.3 The Evolution of Mutation Rates

In the previous section I explored the idea that mutations might be guided in the sense that specific mutations (e.g., for penicillin resistance or for

[7] Reference here to "memory" and "preference" implies no attribution of mentality. Consider the same usage as it applies to the adaptive immune system found in vertebrates. The term "immunological memory" is used to indicate that an initial response to a given pathogen causes an enhanced response to subsequent encounters with that pathogen.

color-matching) occur because they would be useful. It is a separate question whether mutation *rates* increase when organisms are doing badly and decline when organisms are doing well. Instead of asking whether a mutation that causes a blue organism to turn red occurs more frequently when the environment is red, I now am asking whether a gene at one locus will evolve because of its impact on the mutation rates at other loci. In the color-matching experiment, that gene would cause a blue organism that mismatches its environment to raise its probabilities of mutating to red *and* to green but would not do so if the organism matches.

Sturtevant (1937) noted that mutations are almost always deleterious, which led him to wonder why the mutation rates in species aren't all zero. Kimura (1967) suggested a possible answer – that reducing mutation rates imposes a physiological cost. As a simple example, consider the fact that some mutations are caused by solar radiation. If the organisms in a species live in intense sunlight, the sun may cause mutations that make organisms less fit. Suppose that organisms can reduce the frequency of these mutations by growing layers of insulation. Organisms incur a caloric cost if they build and maintain this insulation. As genes evolve that cause more and more insulation to be added, less and less benefit arises from adding another layer. At some point the cost of extra insulation exceeds the benefit that comes from a tiny improvement in fitness. The result is that the mutation rate does not go all the way to zero.[8]

Some biologists have pointed to the fact that mutation rates increase when populations of microbes are starved; they take this to undermine the thesis that mutations are random (see, for example, Fitzgerald and Rosenberg 2019). The idea here is that upping the mutation rate when there is stress is an *adaptation*. Even though most mutations are deleterious, desperate times call for desperate measures. Recall the definition of adaptation discussed in Section 2.7; it says that for a trait to be an adaptation, it must have evolved because there was selection for that trait.

Before embracing the hypothesis that nonzero mutation rates are adaptations, we should consider an alternative explanation for why populations of starving microbes increase their mutation rates. It is suggested by a nice example that George C. Williams (1966, p. 11) presented in *Adaptation and*

[8] The evolution of reduced mutation rate is different from the evolution of improved mutation repair systems. The problem of diminishing returns affects the latter process as well.

Natural Selection. Consider the fact that flying fish return to the water after they glide through the air. Flying fish can't stay aloft forever. Given relevant facts about flying fish, their splashdowns are "mechanically inevitable."[9] Williams says that there is no need to invoke natural selection to explain why flying fish move from air to water and concludes that this movement is not an adaptation. Here Williams is using his idea that adaptation is an "onerous concept" and is deploying the pattern of default reasoning I discussed in Section 5.4.

Fortunately, Williams's conclusion about flying fish can be reached without using the idea that adaptation hypotheses are guilty until proven innocent, and without using the idea that returning to the water is a physical inevitability.[10] The simple point is that adaptations, by definition, are outcomes of selection processes, and selection processes, by definition, require variation. The lineage leading to current flying fish did not include individuals that start their lives in water and then spend the rest of their reproductively successful lives airborne.

How does this idea apply to the question of whether it is an adaptation that microbial populations increase their mutation rates when they are starved? Maybe starvation causes breakdowns in numerous biological processes, high-fidelity gene replication included. If ancestral populations of microbes *always* increased their mutation rates when they starved, there was no variation among those populations in that respect. If so, it is not an adaptation that starving populations now increase their mutation rates. The fact that increased mutation rates sometimes deliver beneficial mutations may simply be a *fortuitous benefit* (a concept discussed in Section 3.6).

If mutation rates are and always have been positive, what is the minimum possible mutation rate? Current species differ in their mutation rates. Whether different species have different physical minima is a good question. Answering it would provide a baseline for determining whether selection has played a role in adjusting mutation rates.[11]

[9] Williams (1966, p. 6) uses another gravitational example to make the same point – "the trajectory of a falling apple."

[10] **Question:** If determinism were true, would that make all traits of all organisms physically inevitable, and would that force Williams to conclude that there are no adaptations?

[11] The importance of having that baseline may remind you of Watson and Galton's discussion of the extinction of family names (Section 5.1).

In addition to this cross-species comparison of mutation rates, differences in mutation rates *within* a species can be evidence that selection has influenced mutation rates. Monroe et al. (2022) report that "mutations [in a species of *Arabidopsis*] occur less often in functionally constrained regions of the genome – mutation frequency is reduced by half inside gene bodies and by two-thirds in essential genes" and that "genes subject to stronger purifying selection have a lower mutation rate."[12] They argue that epigenetic factors are the reason why. If Monroe et al. are right, this is an example in which mutation rates evolved by natural selection. That's interesting, and so is their additional claim that lowered mutation rates in coding regions evolved without *genes* causing the lowered rate.[13]

6.4 Darwin's Gradualism

Although Darwin's gradualism is logically independent of his thesis that "natural selection is the main, but not the exclusive cause" of evolution (Section 1.5), he brings the two together in the following remark at the end of the *Origin*:

> As natural selection acts solely by accumulating slight, successive, favorable variations, it can produce no great or sudden modification; it can act only by very short and slow steps. Hence the canon of "Natura non facit saltum" [nature does not make jumps], which every fresh addition to our knowledge tends to make more strictly correct, is on this theory simply intelligible." (p. 471)

Note that Darwin does not explain in this passage why evolution by natural selection entails gradualism; he just asserts the connection. Note

[12] This finding goes beyond what was already known – that mutation rates vary within genomes; for example, see Wolfe et al. (1989).

[13] Jablonka and Lamb (2005, p. 94) note that it is controversial whether the global mutation rate in a species is an adaptation but claim "there is no doubt that … local hypermutation … is an adaptation." By "local hypermutation," they mean mutations that "occur at a place [in the genome] where they are useful. Certain regions of the genome have a rate of mutation that is hundreds or thousands of times higher than elsewhere … in the jargon of genetics they are 'mutational hot spots.' The genes in these hot spots code for products that are involved in cellular function requiring a lot of diversity. That is what makes the high local mutation rate adaptive." **Question**: how is the distinction between a trait's being *adaptive* and a trait's being an *adaptation* (Section 2.7) relevant to assessing Jablonka and Lamb's thesis?

also that Darwin connects his evolutionary gradualism to a principle that extends far beyond the compass of biology. I mentioned in Section 1.7 the influence of the gradualist geology of Charles Lyell (1830–1833) on Darwin's thinking. Did Darwin think that gradualism in earth history lends credence to gradualism in biological evolution? Did he think that gradualism as opposed to saltationism (sometimes called "catastrophism") is a prerequisite for doing science on any subject? Was Darwin's gradualism a metaphysical article of faith?

Whatever the answers are to these questions, Darwin has a take on gradualism that is specifically biological – it derives from his copious knowledge of artificial selection. Again in the *Origin*, he writes that plant and animal breeders "can largely influence the character of a breed by selecting, in each successive generation, individual differences so slight as to be quite inappreciable by the naked eye. This process of selection has been the great agency in the production of the most distinct and useful domestic breeds (p. 467)." Not only is it true that artificial selection *can* proceed gradually; Darwin knew that it *actually* has done so. Even so, a question arises – why is the fact that breeders have worked this way evidence that natural selection does so too? Maybe natural selection sometimes departs from the gradualism that characterizes artificial selection. In *The Variation of Animals and Plants Under Domestication*, Darwin (1868, p. 414) offers a less monolithic picture of the place of gradualism in the annals of artificial selection. He cites several varieties "that suddenly appeared in nearly the same state as we now see them," although the overwhelming majority "were formed by a slow process of improvement." Darwin (1868, pp. 234–235) explains this frequency difference as follows: "As conspicuous deviations of structure occur rarely, the improvement of each breed is generally the result ... of the selection of slight individual differences." This raises the question of why the rarity of big desirable mutations in artificial selection is a good argument for gradualism in natural selection. Maybe plant and animal breeders are impatient to proceed whereas natural selection is under no such pressure of time.[14] In another passage from the same book, Darwin embraces gradualism more wholeheartedly: "Slight individual differences ... suffice for the work [in both artificial and natural selection],

[14] Alfred Russel Wallace chided Darwin for thinking that artificial selection is a guide to natural selection; see Richard (1997) for discussion.

and are probably the sole differences which are effective in the production of new species" (p. 192). Why does Darwin describe the *sufficiency* of gradualism in both artificial and natural selection and the *probable necessity* of gradualism in speciation? As far as I know, Darwin never explained why he drew this contrast.

In contrast to this bottom-up argument (from observations of artificial selection to gradualism about mindless evolution), there is a theoretical argument for evolutionary gradualism that Darwin could have advanced, but did not. It was constructed by R. A. Fisher in his 1930 book, *The Genetical Theory of Natural Selection*. After presenting his mathematical argument for gradualism based on ideas about random mutation and natural selection (which I'll analyze in the next section), Fisher suggests that his argument accords well with "common experience," and that its intuitive appeal can be perceived by considering an analogy:

> … the mechanical adaptation of an instrument, such as the microscope, when adjusted for distinct vision. If we imagine a derangement of the system by moving a little each of the lenses, either longitudinally or transversely, or by twisting through an angle, by altering the refractive index and transparency of the different components, or the curvature, or the polish of the interfaces, it is sufficiently obvious that any large derangement will have a very small probability of improving the adjustment, while in the case of alterations much less than the smallest of those intentionally effected by the maker or the operator, the chance of improvement should be almost exactly one half (Fisher 1930, p. 44).

Random changes made in a functioning machine are more apt to make a mess of things the larger those changes are. By analogy, random mutations plus natural selection seem to entail gradualism as a thesis about adaptive evolution.

Fisher's microscope analogy raises an interesting historical question. Why didn't Darwin present this analogical argument for gradualism? Instead of discussing a microscope, he could easily have talked about a watch. Darwin (1958, p. 59) says he read William Paley's (1802) book *Natural Theology* while a student at Cambridge and that he was "charmed and convinced" by it. Paley famously used a watch analogy to argue that organisms are products of intelligent design. Darwin rejected intelligent design, but Paley's watch could have served Darwin well in defending evolutionary gradualism. Perhaps Darwin decided not to use this analogy because of its creationist pedigree.[15]

6.5 Fisher's Geometric Argument[16]

Fisher's conception of random mutation (Figure 6.3) differs markedly from Darwin's (Figure 6.1). To get things started, I'll describe what Fisher had in mind by considering a mutation that affects just one phenotype; again my example will be blood pressure. As before, Figure 6.3 represents the organism's present blood pressure as a dot on a line, but now I need to make two changes in the Darwinian picture. First, I'll consider the single optimal blood pressure O, rather than two possible optima. Second, I need to consider a "big" and a "small" mutation. I'll use these terms, not to refer to their absolute sizes, but merely to talk about two mutations where one is bigger than the other. A small mutation has a probability of moving the organism from its dot location to S_1 and the same probability of moving the organism from the dot to S_2; S_1 and S_2 are equidistant from the dot. The same symmetrical arrangement characterizes the big mutation. In this Fisherian picture, the average effect of a mutation on blood pressure is zero, regardless of its size. Fisher assumes that the fitness of a phenotype is a linear function of its distance from the optimum, so the average fitness of a big mutation is the same as the average fitness of a small one. Darwin's Figure 6.1 does not require these symmetry assumptions. Darwin's thesis that mutations are random is compatible with Fisher's, but Darwin's does not entail Fisher's. However, Fisher's entails Darwin's.

The story changes dramatically, Fisher saw, when you move from a mutation that affects one phenotype to a mutation that affects two or more. The case of two is depicted in Figure 6.4. Suppose a mutation affects both blood pressure and blood pH. The symmetry assumptions used in Figure 6.3 apply to Figure 6.4 as well. A mutation, whether it is small or big, will move the organism from its dot location to a point on a circle that is centered on the dot, and all locations on that circle have the same probability (density) of being the mutation's phenotypic result. Whereas the big and

[15] Orr (2009) suggests that Darwin's and Fisher's evolutionary gradualism may have been influenced by their political conservatism.

[16] Material in Sections 6.5–6.7 is drawn from Maxwell and Sober (2023a).

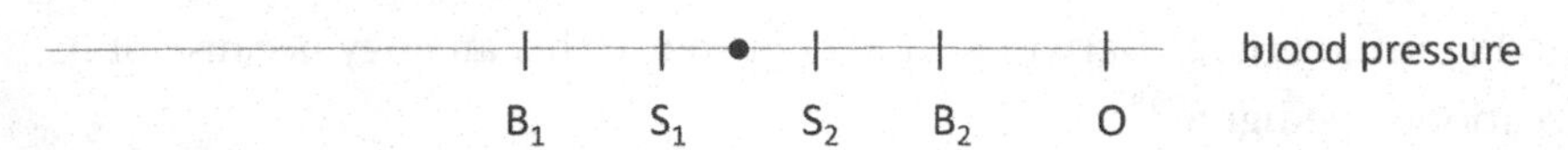

Figure 6.3 Fisher's conception of what "random" mutation means when a mutation affects just one quantitative phenotype.

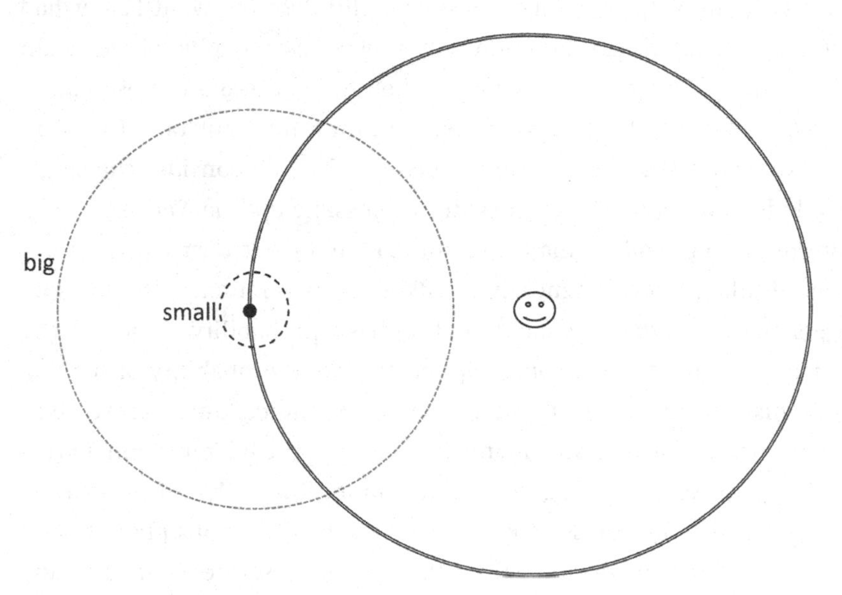

Figure 6.4 Fisher's geometric argument applied to a small mutation and a big mutation, each of which affects two quantitative phenotypes.

small mutations in Figure 6.3 have the same probability of improving fitness, the situation in Figure 6.4 is different. The smiley face represents the optimum. In that figure, the small mutation has a probability of about $\frac{1}{2}$ of improving the organism's fitness, whereas the big mutation has a probability of about $\frac{1}{3}$ of doing so. A random mutation that is even bigger, will have an even smaller probability of improving the organism's fitness. And if the mutation changes the organism's pair of phenotypes to a location that is more than twice the distance from the dot to the optimum, its probability of improving fitness is zero.

Here's a summary of Fisher's argument. If mutations are random in the sense of the symmetries I've described and the fitness of a phenotype is a linear function of how close it is to the optimum, then it follows that

(1) Pr(Mutation M improves the organism's fitness | M is smaller) >
Pr(Mutation M improves the organism's fitness | M is larger),

so long as the smaller mutation induces a change that is less than twice the distance to the optimum. Another consequence is a thesis about the expected (= average) degree of improvement a mutation of given size will induce:

(2) E(improvement in the organism's fitness | M is smaller) >
E(improvement in the organism's fitness | M is larger).

Both expected values in (2) are negative if the mutations change the organism's phenotypes at all; you can see this by examining Figure 6.4. On average, random mutations reduce an organism's fitness, regardless of how small they are. Although (1) and (2) are true, given Fisher's assumptions, the following inequality is not:

E(improvement in the organism's fitness | M is smaller & M improves fitness) >
E(improvement in the organism's fitness | M is larger & M improves fitness).

Figure 6.4 shows why this last inequality is false. Although Fisher's argument as described so far concerns mutations that affect two quantitative phenotypes, the argument generalizes to n phenotypes. Increasing n entails that a mutation of given size has a lower and lower probability of being advantageous.

Does Fisher's model provide a good argument for gradualism?[17] Well, questions can be raised about his assumptions – for example, that mutations are random in the sense depicted in Figure 6.4 and that the fitness of a phenotype is a linear function of its closeness to the optimum. However,

[17] Fisher thought so. He says that "such large mutations occurring in the natural state would be unfavorable to survival, and as soon as the numbers affected attain a certain small proportion in the whole population, an equilibrium must be established in which the rate of elimination is equal to the rate of mutation" (Fisher 1930, p. 41). Fisher's idea is that big mutations are almost always driven to extinction, but that the population continues to have some big mutations present at low frequencies owing to continuing mutational input. Clarke (1971) also takes the geometrical argument to be an argument for gradualism. Kimura (1983, p. 150) and Orr and Coyne (1992, p. 728) argue that gradualism does not follow from Fisher's assumptions.

even if you grant these assumptions, there's still a gap. Inequalities (1) and (2) don't talk about the probabilities of *fixation* of big and small mutations, and this is what matters to the question of whether gradualism is true. That is, gradualism says:

(F1) Pr(Gene G is now at fixation | G was the result of a smaller mutation and G affects phenotypes $P_1, \ldots, P_m$) >
Pr(Gene G is now at fixation | G was the result of a larger mutation and G affects phenotypes $P_1, \ldots, P_m$).

I call this "F1" because it involves *forward* probabilities[18] and the phenotypes are all affected by a *single* gene.

Do propositions (1) and (2) entail F1? The answer is *no*. In fact, what follows from Fisher's assumptions is that F1 is *false*! A bigger mutation often has a higher probability of reaching fixation than a smaller one. My point here is not that a bigger *advantageous* mutation sometimes has a higher fixation probability than a smaller *advantageous* mutation. That's true but irrelevant. What is true is that a bigger mutation often has a higher *average* probability of reaching fixation than a smaller mutation has when you average over all the possible phenotypic upshots of each.

How can that be? As described in Section 5.6, Kimura (1957, 1962) derived a formula (which I called FIX) for a mutation's fixation probability, given its selection coefficient and the effective population size N_e. Figure 6.5 depicts what Kimura's formula says when the effective population size is $N_e = 100$, but the point I'll make is fully general, as the curve would look much the same for much larger values of N_e. Recall from the previous chapter that when a mutation is strictly neutral its chance of fixation is 1/(2N), so in the case at hand the probability of fixation is 1/200 if s = 0. When s is negative the probability of fixation is even smaller. Now consider a bigger and a smaller mutation; each has an average value of s that is negative, where Big's average is more negative than Small's. Big can produce negative s values that are more extreme than the ones that Small can produce, but these more extreme negative s values have a low "cost," in that they can't be much smaller than 1/200 (which is already pretty small). On the other hand, Big can produce more extreme *positive* values than Small, and the probability of fixation goes up rapidly as s becomes positive, which you see

[18] F1 involves forward conditional probabilities, meaning that the conditioning proposition describes an earlier time while the conditioned proposition describes a later.

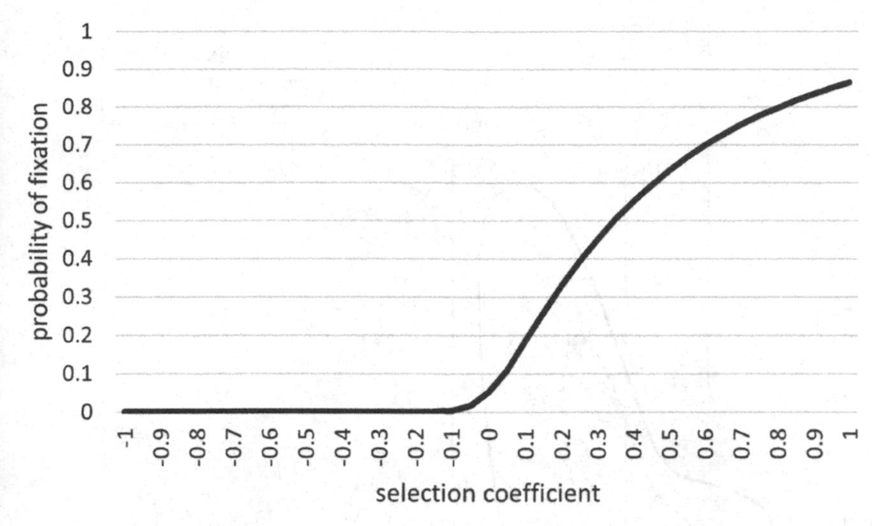

Figure 6.5 Probability of fixation when $N_e = 100$ for different values of s.

in Figure 6.5. This means that the "benefits" that Big might enjoy can be substantial. Small plays it safe while Big is a risk-taker.

Figure 6.6 was constructed by using Figure 6.4 (the one with the smiley face) and adding a third dimension, which is the probability density of fixation given a mutation's upshot for the two phenotypes, where the pair of phenotypes generated by a mutation determines its s value. Here I'm still using Fisher's assumption that the fitness of a phenotype pair is a linear function of how close it is to the optimum. You can see from Figure 6.6 that the small mutation and the big mutation have pretty much the same low probabilities of fixation when they move the organism *away* from the optimum. However, when they move the organism *toward* the optimum, the big mutation picks up larger probabilities of fixation than the small mutation does. In consequence, the big mutation has a higher *average* probability of fixation than the small one. The reason why big has the higher average fixation probability is the nonlinearity of the curve in Figure 6.5.

The question then arises of what the "optimum" phenotypic effect size is for a random mutation, meaning the size that has the highest average probability of reaching fixation. Fisher's assumption that fitness is a linear function of closeness to the optimum entails a surprising answer: when there are two phenotypes that a mutation affects, the optimum size is 73 percent of the distance from the organism's present state to the optimum. That will

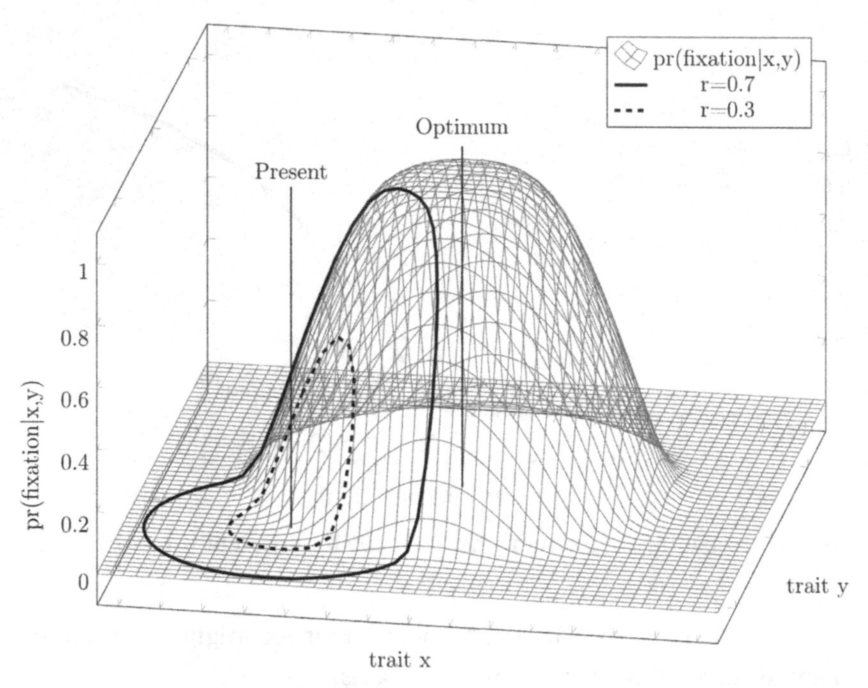

Figure 6.6 Probabilities of fixation for two sizes of mutation ($N_e = 10$).

often be a BIG mutation! Other assumptions about the relationship of phenotypic distance to fitness also entail that bigger mutations often have higher average probabilities of reaching fixation than smaller ones, though the optimal mutation size will be different (Orr 2000; Maxwell and Sober 2023a).

Consider the following argument and assume that the size of the smaller mutation is less than twice the distance to the optimum:

> The average fitness of a smaller mutation is greater than the average fitness of a larger mutation.
>
> When selection is in control of the evolutionary process, fitter traits increase in frequency and less fit traits decline.
>
> ---
>
> Therefore, when selection is in control, the smaller mutation will increase in frequency and the larger mutation will decline.

The first premise is true, and the second premise is the rule of thumb I've used repeatedly. However, the conclusion is false. If the argument is valid, the rule of thumb must be false. On the other hand, maybe the rule of thumb is correct and the argument is simply a fallacy.

I think the second option is the right one. To see why the rule of thumb is not where things go wrong in this argument, consider the fact that Big and Small are not traits that change frequency under natural selection in Fisher's model or in the Kimura correction of that model. When a mutation of any size occurs and is transmitted to the next generation, what is transmitted is the new state of the gene and its associated phenotypes, not the size of the mutation that initially gave rise to that new state. The ensuing selection process "forgets" the size of the mutation once it occurs. The situation here is quite unlike the evolution of traits that are exemplified in succeeding generations.[19]

6.6 From One Mutation to a Sequence of Mutations

Both Fisher and Kimura considered whether a bigger mutation has a higher probability of going to fixation than a smaller one on the assumption that mutations are random in their phenotypic effects. Allen Orr (1998) took the important step of thinking about a *series* of mutations, each affecting the same cluster of phenotypes. If the population is to move from its current set of phenotypes to the optimum, this may involve a sequence of mutations that differ in the size of their phenotypic effects. Orr identified an optimal *sequence* of mutation sizes. Here "optimal" means the sequence of mutation sizes that have the highest probability of going to fixation.

To illustrate Orr's idea, suppose two phenotypes are now at fixation, where the two are caused by the presence of n genes. Each of these genes arose by mutation and then went to fixation. Suppose these mutation/fixation events are spaced out temporally, so that one gene arises by mutation and then goes to fixation, after which another does the same. Using Fisher's linearity assumption, which entails that the optimal mutation size is 73 percent of the distance d to the optimum, the optimal sequence of mutation events that have different phenotypic sizes will be:

(S*) The first mutation's size is (73 percent)d, the second's size is (73 percent)(27 percent)d, the third's is (73 percent)(27 percent)^{2}d, …, and the nth mutation is of size (73 percent)(27 percent)$^{n-1}$d.

[19] **Question:** Does Fisher fall into this fallacy in the passage quoted in footnote 17?

If n is large, the result is a small number of very large mutations followed by a large number of very small ones. S* is an exponential distribution.

Suppose we now look at a particular suite of m phenotypes and n genes that all are at fixation, where each phenotype is affected by all n genes and no others. Does Orr's result entail that those genes *probably* arose by mutations that conform to sequence S*? The answer is *no*. To see why, you need to distinguish these two claims:

(Fn) For each $S \neq S^*$,
Pr(phenotypes $P_1 \ldots P_m$ are now at fixation | genes $G_1 \ldots G_n$ originated by n mutation events that followed S*) >
Pr(phenotypes $P_1 \ldots P_m$ are now at fixation | genes $G_1 \ldots G_n$ originated by n mutation events that followed S)

(Bn) For each $S \neq S^*$,
Pr(genes $G_1 \ldots G_n$ originated by n mutation events that followed S* | phenotypes $P_1 \ldots P_m$ are now at fixation) >
Pr(genes $G_1 \ldots G_n$ originated by n mutation events that followed S | phenotypes $P_1 \ldots P_m$ are now at fixation)

The forward-directed thesis Fn is true; in fact, it simply expresses the idea that S* is optimal. However, Fn does not entail Bn.[20] Fn involves forward-directed probabilities, whereas Bn's probabilities are backward-directed.

You can see why Fn and Bn are logically independent of each other by considering an analogy. A BRCA (pronounced "braca") gene in a human being (female or male) raises the probability of breast cancer, but far fewer than 50 percent of breast cancers trace back to BRCA genes. In other words,

Pr(S now has breast cancer | S had a BRCA gene earlier) >
Pr(S now has breast cancer | S did not have a BRCA gene earlier)

is true, but

Pr(S had a BRCA gene earlier | S now has breast cancer) >
Pr(S did not have a BRCA gene earlier | S now has breast cancer)

[20] **Question:** Does Fn entail F1 when you set n=1?

is false. Notice that the first of these inequalities compares forward-directed conditional probabilities while the second compares probabilities that are backward-directed.[21]

Although Fn does not say that the genes that now encode a phenotype *probably* arose by a series of mutation events that conform to the optimal sequence S*, it does say that the observation of a cluster of phenotypes that is caused by the same n genes *favors* the hypothesis that those genes originated by an S* sequence of mutation events over the hypothesis that it arose by any alternative specific sequence S. Here I use "favoring" in the sense of the Law of Likelihood (Section 4.3).[22]

What does Fn say about gradualism? Sequence S* is more gradualist than many alternative sequences, but S* is less gradualistic than many others. In fact, there are infinitely many alternative S sequences in both categories. Rather than trying to shoehorn Fn into one of two categories (extreme gradualism and extreme saltationism), it is better to simply recognize that sequence S* is an optimal mixture of bigger and smaller. That optimum will include a few large mutations and a lot of smaller ones when n is large, and so sequence S* is "closer" to extreme gradualism than it is to extreme saltationism.[23]

The Fn proposition leaves open how often the assumptions that go into its derivation are satisfied in the real world. The argument for Fn relies on the assumption of pleiotropy. How often do single genes have multiple phenotypic effects? Another assumption is that the optimum does not change as the population evolves. If the phenotypic optimum moves away from the organism's phenotypic location as fast as the population evolves towards that moving target, the optimal sequence of mutation sizes will be a series of large phenotypic jumps. This is the situation envisaged by van Valen's (1973) Red Queen hypothesis (discussed in Section 2.8).

[21] **Exercise:** Show how Fn and Bn are connected to each other by using the odds formulation of Bayes's theorem discussed in Chapter 4. If Fn is true, what else do you need to know in order to decide whether Bn is true?

[22] Moutinho et al. (2022) describe evidence for Orr's "adaptive walk" hypothesis drawn from *Drosophila* and *Arabidopsis*.

[23] On the other hand, if a population reaches the optimum by instantiating Orr's optimal sequence of mutation sizes, the lion's share of the distance traveled was achieved by bigger mutations going to fixation, and this might be said to accord more with saltationism than with gradualism. I thank Kiel McElroy for this point.

6.7 Genome-Wide Association Studies

Can genome-wide association studies (GWASs) help answer the question of whether genes that now are at or near fixation usually trace back to mutations that had small phenotypic effects? These studies examine different polymorphic loci and tabulate how much phenotypic difference there is among the genotypes at a locus while controlling for the genotypes present at other loci as much as possible. GWAS papers often report that phenotypic effect sizes are usually "small."[24] Is this an argument for gradualism? Three cautions are in order. First, GWASs are often about disease phenotypes, and therefore may constitute a biased sample. Second, phenotypes are caused by genes *and environments*, and it's possible that present genes now have smaller effects on phenotypes than their precursors did in earlier environments. But most centrally, it is important to recognize that the theoretical arguments of Fisher, Kimura, and Orr are about mutation sizes, whereas GWASs aren't about that at all. To see why, consider a simple example. Suppose there are two alleles, A_1 and A_2, at a locus in a population and they simultaneously mutate to B_1 and B_2, respectively, after which B_1 replaces A_1 and B_2 replaces A_2. This means that there were three genotypes before and three genotypes after. Suppose the average length of a bone in individuals with these six genotypes were as follows:

$$H(A_1A_1) = 3.0 \text{ cm}, H(A_1A_2) = 3.01 \text{ cm}, H(A_2A_2) = 3.02 \text{ cm}$$
$$H(B_1B_1) = 6.0 \text{ cm}, H(B_1B_2) = 6.01 \text{ cm}, H(B_2B_2) = 6.02 \text{ cm}$$

The mutations are big, but the average effect sizes at the locus, before and after the mutations occur, are small.

6.8 Concluding Comments

Mutation rates vary among present day species, but that does not entail that mutation rates have evolved, even given the fact of universal common ancestry. Mutation rates could have changed historically simply because

[24] See, for example, Park et al. (2010), Visscher et al. (2010), and Simons et al. (2018). Marouli et al. (2017) did a GWAS on human adult height and report that "variants with a larger effect size on height variation tend to be rarer."

mutagens in the environment changed. Evolution means something more than change (Section 1.3). Similarly, the hypothesis that mutation rates have evolved does not entail that they evolved because of natural selection. And the hypothesis that mutation rates have evolved by natural selection does not show that organisms have the ability to adaptively adjust their mutation rates, nor that they have the ability to cause specific beneficial mutations to occur when needed.

Darwin's idea that mutations are random in the sense of Figure 6.1 still looks good, but Fisher's argument for gradualism does not. Fisher's argument is about the average effect of mutations on the fitnesses of the organisms that experience them, and it therefore does not answer the question of how the size of a mutation affects its fixation probability. This is where Kimura's formula comes into play and undermines the Fisherian argument for gradualism. Orr reformulated the question about mutation size, shifting from the earlier question about a single mutation to a new question about a sequence of mutations, all affecting the same cluster of phenotypes. His argument leads to a formulation of "tempered" gradualism (in which a few mutations have large effect and many have small) that has a likelihood justification (which does not mean that that formulation is probably true!). Crucial to his argument is the assumption that populations evolve towards stationary optima.

I mentioned at the start of the previous chapter that Darwin recognized the possibility that traits that have no effect on an organism's survival and reproduction can evolve "by chance" (or as we now say, by drift). He did not rule this out even though he insisted that selection is the main cause of evolution. Perhaps Darwin should have taken the same relaxed attitude towards gradualism. This is what Thomas Henry Huxley (famous for defending Darwin's theory against Bishop Wilberforce's attack in an Oxford debate in 1860) suggested in his favorable review of the *Origin*:

> Mr. Darwin's position might … have been even stronger than it is if he had not embarrassed himself with the aphorism, 'Natura non facit saltum,' which turns up so often in his pages. We believe … that Nature does make jumps now and then, and a recognition of the fact is of no small importance in disposing of many minor objections to the doctrine of transmutation (Huxley 1860).

In the light of Kimura's and Orr's work, Darwin's monolithic insistence that adaptive evolution is a gradual process may need to be replaced by a pluralistic picture in which genes that reach fixation or near-fixation often trace back to a mixture of mutation events that differed in size. Darwin was wise not to take Asa Gray's advice, but perhaps he should have taken Huxley's.

7 Taxa and Genealogy

7.1 Darwin's Hope

Darwin hoped his theory of evolution would put biological taxonomy (aka *classification* or *systematics*) on a new and better footing. Using the idea of universal common ancestry supplemented with the idea that genealogies have the tree-like structure depicted in Figure 1.1, Darwin (1859, p. 420) asserts that "all true classification is genealogical." He contrasts this thesis with alternatives that were already on the table:

> That community of descent is the hidden bond which naturalists have been unconsciously seeking, and not some unknown plan of creation, or the enunciation of general propositions, and the mere putting together and separating objects more or less alike.[1]

Twelve years later, Darwin reaffirms this thesis in *The Descent of Man*, noting that it had by then achieved wide acceptance:

> [N]aturalists have long felt a profound conviction that there is a natural system. This system, it is now generally admitted, must be, as far as possible, genealogical in arrangement—that is the codescendants of the same form must be kept together in one group, apart from the co-descendants of any other form; but if the parent forms are related, so will their descendants, and the two groups together will form a larger group (Darwin 1871, p. 181).

Does Darwin's idea that classification should be genealogical mean that genealogical relationships must *determine* what the taxa are in your classification, or merely that genealogy is an important consideration? That's

[1] This remark of Darwin's reminds me of Carnap's concept of *explication*, which I mentioned in the Preface.

an interesting historical question about Darwin,[2] but my initial focus in this chapter will be on how the relationship of genealogy and taxonomy *should* be understood, after which I'll turn to the topic of genealogy taken by itself.

7.2 Ranks and Hierarchy

A biological taxonomy is a classification of organisms into different taxa. In the influential tenth edition of his book *The System of Nature*, Carl Linnaeus (1758) places the numerous taxa he identifies into one of the following *ranks* – species, genera, families, orders, classes, phyla, and kingdoms. Species have the lowest rank and kingdoms have the highest. As mentioned in Section 1.2, Darwin and others talked about varieties, which are sub-specific taxa. Modern biologists now add a rank above that of kingdoms; there are three *domains* – Eukaryotes, Bacteria, and Archaea. Biologists also have inserted numerous taxonomic ranks between Linnaeus's lowest and highest.

Darwin thought that biological classification should be *hierarchical*; this means that for any two taxa you please, either one of them is nested in the other or they are disjoint – there are no partial overlaps.[3] For example, Mammalia is nested inside of Vertebrata, and Mammalia is disjoint from Aves (=birds). But why should classification be hierarchical? The answer is not that classifications, by definition, must be structured in this way.[4] An answer that appealed to Darwin is that biological classification must be hierarchical because genealogy is tree-like, not reticulate (Section 1.1). However, there can be other rationales for hierarchy; Aristotle's taxonomy was

[2] Ernst Mayr (1982) argued for the second interpretation of what Darwin meant, but Kevin Padian (1999) and others convincingly argue for the first.

[3] This entails that if two organisms are in taxon T, then they are in all the same taxa of higher rank than T.

[4] The periodic table of the chemical elements (first proposed by Dimitri Mendeleev in 1869 and subsequently refined) is not hierarchical – it is periodic. In this classification, elements are organized in a table that has four rows. As you read left to right and top to bottom, as you are doing now in reading the page before you, the atomic number (the number of protons in an element's nucleus) increases by one at each step. The four rows are called periods, while the columns are called groups. Elements in the same group have similar chemical characteristics. Similarity of atomic number and similarity within columns cross-classify each other. See Scerri (2007) for details.

hierarchical but non-evolutionary, and biologists in the twentieth century sometimes asserted that phylogeny and classification are both hierarchical while denying that genealogical relationships determine classification.

7.3 Three Systematic Philosophies

The idea that science and philosophy are entirely separate from each other is belied by numerous cases in the history of science in which scientists argued with other scientists about philosophical questions that mattered to the science they were trying to develop.[5] A good example is the debate among biologists over which of three competing systematic philosophies should be accepted. *Evolutionary taxonomy* was the dominant philosophy at the time of the Modern Synthesis (1930–1960), *phenetic taxonomy* was developed and had its heyday in the 1960s, and *cladistics* came along around the same time.

In Chapter 4, I discussed the idea that observing that individuals A and B have a given trait provides *evidence* that they have a common ancestor. This suggests that if you observe that individuals A and B are more similar to each other than either is to C, that this is *evidence* that A and B are more closely related genealogically than either is to C. This suggestion concerns the epistemology of propinquity of descent. My concern right now isn't epistemological. Suppose you *know* that A and B are more closely related to each other than either is to C. The question of present interest is how that genealogical fact bears on how you classify A, B, and C. I'll get to the epistemology later.

Cladistics, the taxonomic philosophy inspired by the work of Willi Hennig (1950/1966), asserts that genealogical relatedness is the *only* relevant consideration in deciding which groupings of organisms are taxa and which are not. "Clade" is the Greek word for branch. A cladist looking at the phylogenetic trees depicted in Figure 7.1 will tell you that the taxonomic groups are as follows:

- (Lizards, Crocs, Birds), (Crocodiles, Birds), Lizards, Crocodiles, Birds
- (Marsupial Wolves, Placental Wolves, Moles), (Placental Wolves, Moles), Marsupial Wolves, Placental Wolves, Moles.

[5] Recall the discussion in Chapters 4 and 5 of the debate between Bayesians and frequentists about proper procedures in scientific inference.

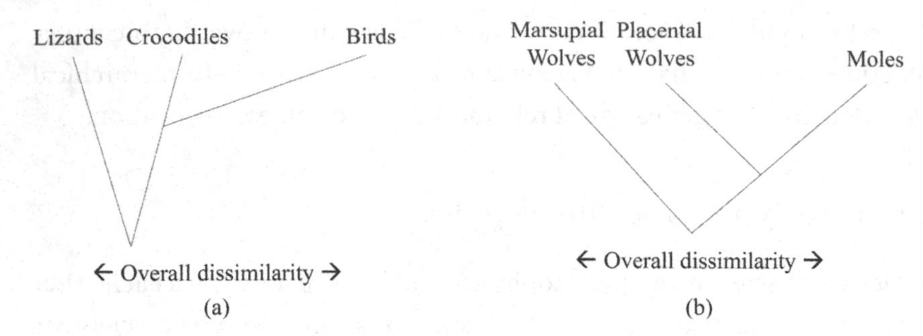

Figure 7.1 Two cases in which genealogy and similarity conflict.

There are ten monophyletic groups on this list. A *monophyletic group* is defined as an ancestor and all and only its descendants. A node in a phylogenetic tree – whether it is at the root or in the tree's interior – represents an ancestor. Lizards + Crocodiles is not a monophyletic group, and neither is Marsupial Wolves + Placental Wolves.[6]

All the monophyletic groups just listed include more than one species; they are "super-specific." However, it doesn't matter to cladists what *rank* these different monophyletic groups are assigned, so long as more inclusive groups get higher ranks than groups that are less. The *ranking* of taxa isn't dictated by genealogy. The two other systematic philosophies I'll now describe agree that super-specific ranking is conventional. Species raise some special questions, which I'll discuss later.

Although cladistics requires super-specific taxa to be monophyletic groups, this does not entail that every monophyletic group is a taxon. A single sexual organism and all of its descendants is a monophyletic group, but if your classification is to be hierarchical, you had better not include such monophyletic groups in your taxonomy. The descendants of one sexual organism and the descendants of another can partially overlap.

The diametric opposite of cladism is *pheneticism*, which says that classification should be based purely on overall similarity. Pheneticism isn't the thesis that similarities are *evidence* that two or more organisms belong to the same taxon. What pheneticists claim is that similarities *define* biological

[6] A word of clarification: if A and B are more closely related to each other than either is to C, then *there exists* a monophyletic group that includes A and B, and *there does not exist* a monophyletic group that includes A and C without including B. There is no implication that *every* monophyletic group that includes A also includes B.

taxa. The distinction drawn in Section 2.2 between "x is evidence for y" and "x defines y" is relevant here. Critics maintain that pheneticism is *operationalist* in character and take that to be a point against it (Hull 1968).

Sokal and Sneath (1963) and other pheneticists think that biological classifications should be "all-purpose." Classifications should be useful to evolutionary biologists, but they also should be useful to biologists who aren't interested in evolution, and to recreational bird watchers as well. Theory-neutrality is required if taxonomies are to be all-purpose. Another reason for theory-neutrality is that pheneticists want their taxonomies to be useful in testing evolutionary hypotheses; they think that usefulness would be compromised if taxonomies were constructed on the basis of evolutionary assumptions (Hull 1970).

The third taxonomic philosophy, *evolutionary taxonomy*, is conceptually "intermediate" between cladism and pheneticism. According to this approach, genealogy trumps similarity in some cases but the reverse is true in others. In considering Figure 7.1a, Mayr (1942), Simpson (1945) and other evolutionary taxonomists would point out that lizards and crocodiles resemble each other because *they retained the traits of their most recent common ancestor* whereas birds diverged from their reptilian ancestors by evolving novel adaptations. Shared traits that satisfy the italicized phrase are called *homologies*. The situation is different in Figure 7.1b. Traits shared by marsupial and placental wolves but not by moles were not present in their most recent common ancestor. The shared states of the two wolves are not homologies; they are *homoplasies* (aka *convergences*). According to evolutionary taxonomy, Lizards + Crocodiles is a taxon ("Reptilia") that excludes birds, but Marsupial Wolves + Placental Wolves is not a group that excludes moles. Homology and homoplasy are concepts that involve genealogy, so one can understand why Mayr and other evolutionary taxonomists thought they were following Darwin's advice that taxonomy must be genealogical.[7]

[7] I've described homologies and homoplasies by talking about traits without distinguishing genotypes from phenotypes, but bringing in that distinction introduces a complication. Suppose two descendants and their most recent common ancestor all have phenotype P and P was retained in the two lineages. Now imagine that one descendant's P is caused by gene (or gene complex) G1 while the other is caused by G2. It is clear enough that G1 and G2 aren't homologous because they don't even match. **Question:** Is the shared phenotype P properly called a homology?

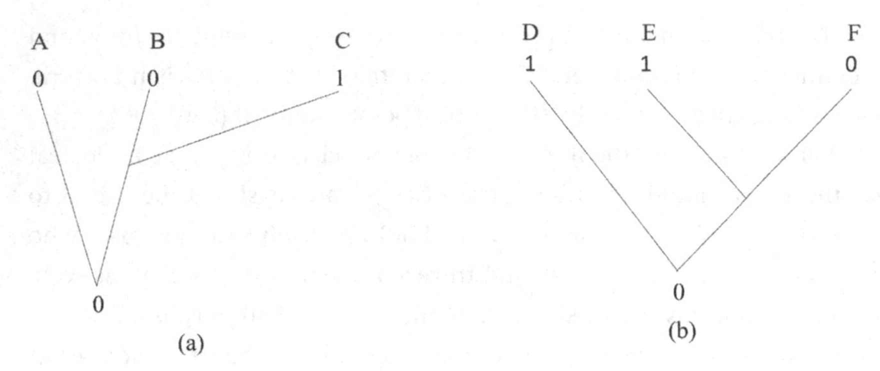

Figure 7.2 Homology and homoplasy.

The difference between Figure 7.1a and b is represented more abstractly in Figure 7.2.[8] Consider a trait whose two states are 0 and 1, where 0 represents the ancestral condition and 1 the derived condition. A and B are in state 0. Is that a homology? Well, it *may be*. However, it is possible that the lineages leading from the root of the tree to A and B changed from 0 to 1 and back to 0, in which case the shared state of A and B is a homoplasy. In contrast, D and E are in the derived state 1 for the character considered, and so their resemblance *must be* a homoplasy.

I've just described the three taxonomic philosophies in what I think is a logical order. Now I want to briefly comment on their chronology relationship. Evolutionary taxonomists appealed to their intuitive judgments and their deep familiarity with the organisms under study to justify their taxonomic groupings; they often emphasized that classifying is an art, not a mechanical procedure. In forming their classifications, evolutionary taxonomists often developed adaptive stories about how traits evolved. Pheneticists disapproved of both the appeals to intuition and the telling of adaptive stories; they wanted the rules for constructing classifications from observations to be explicit and objective, and to exclude speculations about history. The advent of computers allowed them to construct algorithms for constructing taxonomic classifications from observations; these mechanical procedures were anathema to evolutionary taxonomists. Cladists agreed with pheneticists that taxonomic inference needs to be made more

[8] Figure 7.2 departs from 7.1 in another way. In Figure 7.1, the horizontal dimension represents dissimilarity; in Figure 7.2, it does not. Figure 7.2 is a pure "tree topology" – the branching structure represents propinquity of descent and nothing more.

rigorous, and that adaptive story-telling should be avoided. They parted ways over the question of whether classifications should be all-purpose or strictly genealogical.

Here's a criticism of evolutionary taxonomy: it makes classification weirdly *disjunctive*. The reason that A and B are put together apart from C in evolutionary taxonomy's (AB)C classification is entirely different from the reason that E and F are put together apart from D in that taxonomy's D(EF) classification. A and B are put together because their shared traits are thought to be homologies, whereas E and F are put together because of propinquity of descent, even if there are no homologies shared by E and F.

Here's the first of my three criticisms of pheneticism: it groups by overall similarity, with each similarity in one's data set assigned the same weight (Sokal and Sneath 1963; Sneath and Sokal 1973). Equal weighting is often defended by saying that it is unbiased and assumption-free, but equal weighting is just as substantive an assumption as unequal weighting.[9] Pheneticists think that equal weighting is needed because their goal is to provide all-purpose taxonomies. Yet, the divergent interests of bird watchers and evolutionary biologists suggest that a taxonomy aimed at serving both of these audiences will serve neither. Bird watchers often want a guide that classifies by observable features and identifies the species to which the bird you're observing belongs. In contrast, evolutionary biologists often want phylogenies, but at other times they want information about which taxa have a single trait of interest; for example, biologists who study sex ratio evolution need information about the different sex ratios that different species have. Facts about *overall* similarity won't be useful to either of these constituencies.

My second criticism of pheneticism is that the concept of "overall similarity" admits of several interpretations, even after a set of traits is selected and

[9] This is analogous to a standard challenge that *the principle of indifference* faces in probability theory. The principle says that if $P_1, P_2, \ldots, P_n$ are exclusive and exhaustive propositions, and you have no reason to think that they are unequal in probability, you should assume that they are equally probable. The criticism of the principle is that assuming that probabilities are equal is just as substantive as assuming that they are unequal. It is misleading to call the equiprobable distribution an "ignorance prior." Believing that the probabilities are equal is different from not knowing what to believe. This point leaves open whether there is a good defense of the principle of indifference that addresses this challenge.

Table 7.1 *Phenetic groupings differ depending on which trait is used*

		Organisms			
		X	Y	Z	Phenetic Grouping
Traits	Length	1	2	3	XYZ
	$Length^2$	1	4	9	(XY)Z
	$\sqrt{Length}$	1	1.41	1.73	X(YZ)

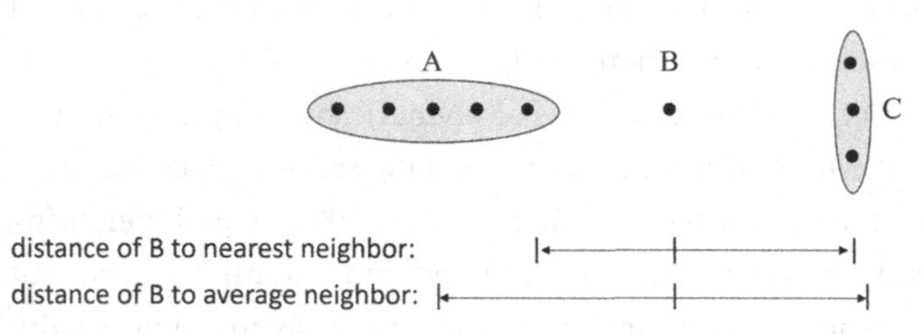

Figure 7.3 Nearest neighbor versus average neighbor clustering.

they are assigned equal weight. The problem is illustrated in Figure 7.3.[10] Suppose you've scored each of the dot objects for two quantitative features, one represented on the x axis, the other on the y (as I did in Figure 6.4 in connection with Fisher's geometric argument). The A objects form one cluster and the C objects form another. To which of these clusters does B belong? If you put B with its nearest neighbor, you'll put it in A; if you put B with the cluster whose average member is closer, you'll put it in C.

Table 7.1 illustrates my third and final criticism of pheneticism, which I think may be the most fundamental. Suppose the tibia lengths in organisms X, Y, and Z are one foot, two foot, and three foot, respectively. Since Y is halfway between X and Z, it is arbitrary to put Y with one of them rather than the other, and anti-phenetic to put X and Z in a group apart from Y. So pheneticists will take tibia length to indicate that no two of X, Y, and Z should be put in a group apart from the third. Matters change if you look at the squared length or the square root of the length, as shown in the table. Recall that pheneticism says that *all* observed traits are to be considered in classifying and they are to be given equal weight. An organism's value for one way of measuring tibias logically entails its value for the other two, so

[10] This idea is from Johnson (1970) by way of Ridley (1986, pp. 41–42).

they are not logically independent of each other. This means that a choice must be made. By eschewing the use of scientific theories, pheneticism cannot tell you how to choose.[11]

Darwin hoped that the genealogical approach to classification would revolutionize how classifications are understood, but he seems to have expected the new approach to retain almost all of the taxa that his predecessors had identified. In fact, the cladistic revolution produced lots of revisions. Animals, flowering plants, birds, insects, and mammals are monophyletic, but reptiles, invertebrates, moths, and algae are not. Cladistics is now the dominant approach for classifying Eukaryotes, though it is less hegemonic in connection with Bacteria and Archaea. This is because horizontal gene transfer is thought to play a relatively minor role in the former, but a relatively major role in the latter. I will concentrate on cladistics in what follows.

7.4 Taxa Are Tokens, Not Types

Linnaeus's three kingdoms were animal, vegetable, and mineral. For him and many other taxonomists pre-Darwin, mammals are a kind of animal in the same sense that gold is a kind of mineral. This raised no eyebrows so long as taxa were thought of as *natural kinds*. Cladistics changed all that. The standard interpretation of cladistic taxa is that they are token physical objects (Section 2.5) that have organisms as their parts. These parts are spread out in time and space, and the time slices are connected to each other by reproductive relationships. Monophyletic groups originate with an ancestor, and they go extinct when no descendants of that ancestor exist. They resemble token organisms that are born and die.

[11] This point about similarity resembles an example that Van Fraassen (1989, p. 303) gives to illustrate the Bertrand paradox, which concerns the principle of indifference in probability theory. Van Fraassen describes a factory that produces cubes that are always less than or equal to 2 cm on an edge. Given just this information, what is the probability that a cube made by the factory will have an edge length that is less than or equal to 1 cm? You might say that the right answer is that the probability is ½, but now consider a second question. Given that the factory always produces cubes that are at most 4 cm^2 on a side, what is the probability that a cube made by the factory will have a side whose area is less than or equal to 1 cm^2? If you say 1/4 here, your two sayings contradict each other. The principle of indifference can't be applied both to lengths and areas. You need to choose, and the choice can't be justified a priori. You need to look at the factory's production process.

You are not a kind (though you may be one of a kind); the same point holds of the mammal clade. Natural kinds are different. Token objects count as gold because of the features they share; they need not have a common ancestor.

It isn't just that taxa are tokens rather than types. The additional point is that the status of taxa as monophyletic groups means that all the usual traits that were supposed to "define" taxa turn out to be inessential. Consider vertebrata as an example. It used to be said that a vertebrate is an organism that has a spinal cord, and that this is true by definition. However, if vertebrata is a monophyletic group, then *all* the descendants of vertebrate organisms are vertebrates, even (hypothetical) descendants that do not have spinal cords. What is arguable is that the *first* vertebrates had spinal cords. This is *the most parsimonious estimate* of the character states of those ancestors, given the observation that all present-day vertebrates have spinal cords; I'll discuss parsimony later in this chapter.

The idea that species and other taxa are tokens, not types, has often been formulated by saying that they are *individuals* (Ghiselin 1974; Hull 1978). This choice of words has led to confusion, since "the evolution of individuality" has been an important topic in evolutionary theory (Buss 1987), but it's different from the point about type and token. Vertebrate organisms have a high degree of individuality, meaning that the parts of those organisms exhibit a high degree of physiological dependence. Cut away 30 percent of a tiger and the tiger will die. The same isn't true of a grove of aspens. The trees are connected to each other by underground runners, but severing these runners doesn't result in the trees dying (Sober 1991b).

If individuality is the quantitative feature just described, it is easy to see why the thesis that species are individuals is jarring. The organisms in a cosmopolitan species depend on each other physiologically far less than do the cells of a tiger. For super-specific taxa, the level of interdependence is often even less. It is for this reason that Wiley (1981) suggested that the term "historical entities" be used instead of "individuals" to make the point I've described concerning type and token.

7.5 Essentialism

Ernst Mayr (1959) and David Hull (1965) argued that pre-Darwinian taxonomy was hostage to the pernicious philosophical doctrine of essentialism espoused by Aristotle. The title of David Hull's paper sums it up – "The Effect of Essentialism on Taxonomy – 2000 Years of Stasis." This historical thesis aside,[12] I want to clarify what essentialism is and how it is related to Darwinian evolution.

Gold is a natural kind and it has an essence. The essence of gold is a property that all and only the gold things possess. The possession of that essence is what makes each of those things gold,[13] and it explains many of the observable features that samples of gold possess. This is what the doctrine of *kind essentialism* demands. According to essentialism, the essence of gold is not something mysterious and unscientific. It was discovered by chemistry.[14]

Even if essentialism is fine for chemistry, does it make sense for biological taxonomy? David Hull's (1965) argues as follows that the answer is *no*:

The essence of a natural kind never changes.
If species are natural kinds with essences, they cannot evolve into new species.
Species do evolve into new species.

Species essentialism is false.

The idea behind the first premise is that the *membership* of a natural kind can change (e.g., gold wedding rings can be created and destroyed), but *the kind itself* is always the same – the essence of gold was and always will be atomic number 79. However, once the first premise is understood in this way, you can see why the second premise is false. Just as an atom smasher

[12] Mayr and Hull thought that Aristotelian essentialism dominated taxonomy until Darwin. Winsor (2006a, 2006b) and Müller-Wille (2007) argue that Linnaeus was not an essentialist and that the Mayr–Hull narrative is wrong. Lennox (2001) argues that Aristotle's essentialism was teleological, not typological; the "natural state model" that Gottlieb and Sober (2017) attribute to Aristotle entails that Aristotle's essentialism was both. Lennox seems to be thinking about observable phenotypes when he denies that Aristotle was a typological essentialist, not about the more "theoretical" construct of a natural state.

[13] My phrase "makes it the case" does not denote a causal relationship; it describes a synchronic, not a diachronic, relation. A puddle of water's being composed of H_2O molecules at time t is what makes it the case that the puddle is made of water at time t.

[14] Sidelle (1989, 2009) argues that there is a conventional element in the thesis that gold is a natural kind that has an essence.

might be able to transmute lead atoms into gold, so evolution might have the result that a lineage with a given suite of traits at one time has a very different suite later on, with the result that ancestors and descendants belong to different species.

Another failed argument against taxonomic essentialism begins with the assertion that essentialism precludes gradual evolution. Gold, so it is said, has an essence, which provides a crisp necessary and sufficient condition for being gold. An atom either has atomic number 79 or it does not. In contrast, present-day organisms and their ancient ancestors may belong to different species even though there is no bright line on the lineage connecting the two that marks where the old species stops and the new one begins. My reply is that the vagueness involved in the process of gradual evolution does not compromise the thesis that taxa are natural kinds and have essences. In the hypothetical transmutation of a lump of lead into a lump of gold, there might be intermediate stages in which it is indeterminate whether the lump is lead or gold.[15]

The thesis that taxa have essential properties is usually associated with the idea that taxa are natural kinds, but this connection isn't inevitable. For example, consider Saul Kripke's (1972/1980) thesis of *origin essentialism*, which holds that one of your essential properties is that you have the very parents you had. Had your parents not existed, you would not have existed. This thesis is compatible with the fact that an individual just like you could have existed even if your parents had not, but that doppelganger would not have been you. The thesis of origin essentialism can be applied to cladistics; the result is that a monophyletic group comprised of ancestor A and all its descendants would not have existed if A had failed to exist. True, a clade that looks just like Aves could have existed without the existence of the token ancestor that defines that monophyletic group, but it would not have been the same token taxon, or so the thesis of origin essentialism maintains (Sober 1980a, Griffiths 1999, Okasha 2002, LaPorte 2004).[16]

[15] Devitt (2008) argues that "Linnaean taxa, including species, have essences that are, at least partly, underlying intrinsic, mostly genetic, properties." Boyd's (1999) idea that species and higher taxa are "homeostatic property clusters" also treats taxa as natural kinds.

[16] Ereshefsky (2014) and Pedroso (2014) argue against applying origin essentialism to species.

Origin essentialism and kind essentialism are different. As noted, the former pertains to token objects while the latter pertains to types, but there is a further difference. Kind essentialism requires that kinds have essences that play an important role in explaining the other traits that items of that kind possess. Origin essentialism makes no such demand. The claim that the descendants of ancestor A would not have existed if A had failed to exist does not require that A's existence explains the traits of A's descendants (Okasha 2002; Ereshefsky 2022).

7.6 The Taxonomic Status of Ancestors

Using the concept of monophyly to define what a taxon is has two surprising consequences. To see the first, let's return to the fact that for any two monophyletic groups in a phylogenetic tree, one is part of the other or the two are entirely disjoint. If one taxon is part of another or if the two are disjoint, neither is ancestral to the other. Therefore, no monophyletic group is an ancestor of another. This entails that no taxon is an ancestor of another if all taxa are monophyletic.[17]

This conclusion leaves room for a taxon's having *past organisms* or *past populations of organisms* as ancestors, and those ancestors can belong to taxa.[18] Our species has organism ancestors that were not members of our species; some of them were non-human mammals, while others were non-mammalian vertebrates. Can those ancestors belong to a species distinct from our own? They cannot, if species are monophyletic. This point generalizes to super-specific taxa. The organism ancestors of a taxon of rank X belong only to taxa of higher rank.

This means that the objects you put on the leaves of a phylogenetic tree have implications about the taxonomic categories to which the interior nodes of the tree can belong. If you put whole monophyletic taxa on the leaves of a phylogenetic tree, the interior nodes will belong only to taxa of higher rank. On the other hand, if you construct a tree on which the leaves are current organisms or populations, the ancestors represented by

[17] **Exercise:** How can this argument be generalized to take account of the possibility that genealogies can be reticulate?

[18] Here we need to distinguish a node in a phylogenetic tree *belonging to* monophyletic group M and its being *identical with* monophyletic group M.

interior nodes can belong to monophyletic taxa at all ranks. It would be misleading to put this point by saying that only current organisms or populations can occupy the leaves of a phylogenetic tree.[19]

7.7 Species

Cladists require all taxa to be monophyletic, but non-cladists have no such commitment. Pheneticists, for example, hold that *all* taxa, species included, are groups of individuals defined by overall similarity. The individuals in a sub-specific variety are more similar to each other than are the individuals in the species to which that variety belongs, and the individuals in a species are more similar to each other than are the individuals in the genus to which that species belongs.[20]

This pheneticist view of species is at odds with one of the most popular species concepts in biology, namely Ernst Mayr's (1963) biological species concept, which says that "species are groups of interbreeding natural populations that are reproductively isolated from other such groups." It is interesting that this definition describes how *populations* must be related to each other to be conspecific; it does not describe what it takes for two *organisms* to belong to the same species. This is no accident, in that two individuals can be conspecific even though they can't produce viable fertile offspring; think about organisms of the same sex or of sterile individuals.[21]

When Mayr speaks of interbreeding populations, he seems to be talking about populations that *actually* interbreed; this goes beyond the idea that they have the *potential* to interbreed. If so, his definition needs fine-tuning, since populations that exist at one time don't interbreed with those that existed 100 generations earlier or later even if they are genetically and phenotypically identical. A natural adjustment is to say which populations that exist in the same chunk of time are members of the same species, and

[19] For further discussion, see Baum (1998, 2009), Mishler and Wilkins (2008), and Velasco (2008).

[20] For a useful survey of numerous modern species concepts, see De Queiroz (2005), Malet (2007), the appendix of Coyne and Orr (2004), or Ereshefsky (2022).

[21] This feature of Mayr's species concept accords well with his idea that the Darwinian theory of evolution implements a conceptual device that Mayr (1959) calls "population thinking." See Sober (1980a) for discussion.

then stack different generational time slices together by their genealogical connections to form a temporally extended species.

Notice that the biological species concept entails that there can't be such a thing as an asexual species. Many biologists are prepared to bite this bullet, since the process of speciation, which they define as the process by which populations become reproductively isolated from each other, plays an important role in evolutionary biology. The biological species concept describes the product of that process (Coyne and Orr 2004).[22]

Reproductive isolation is an important property in evolutionary biology, but so is the property of monophyly. Biologists often want to describe species in terms of the first property but they also want to put species on the leaves of the phylogenetic trees they construct. These two endeavors can come into conflict (Velasco 2008, Baum and Smith 2013). Present-day Humans and Chimpanzees are more closely related to each other than either is to present-day Gorillas. This means that those Humans and Gorillas must have precisely the same degree of genealogical relatedness to each other that those Chimpanzees and Gorillas have to each other. In just the same way, two siblings have the same degree of relatedness to their cousin. This simple fact entails that it makes no sense to ask how the "group" of present-day Humans+Gorillas is related to present-day Chimpanzees. Present-day Humans and Chimpanzees are closely related to each other while present-day Gorillas and Chimpanzees are related to each other more distantly. There is no such thing as *the* degree of relatedness that Humans+Gorillas bear to chimpanzees.

This point bears on the biological species concept. Figure 7.4 shows three contemporary populations A_1, A_2, and A_3 that belong to the same species as defined by the biological species concept. The branches indicate both lineages and migrations. At the root of the tree there was just one population. It split in two, and then after several generations one of the resulting populations split again. The result is that the individuals in A_1 and the individuals in A_2 are more closely related to each other than either is to the individuals in A_3, but migration among these three populations happens occasionally and the result is reproduction; the three populations are part of a single species as far as the biological species concept is concerned. Now suppose that A_2 evolves a trait that prevents it from reproducing with

[22] For an interesting discussion of species concepts for prokaryotes, see Ereshefsky (2010).

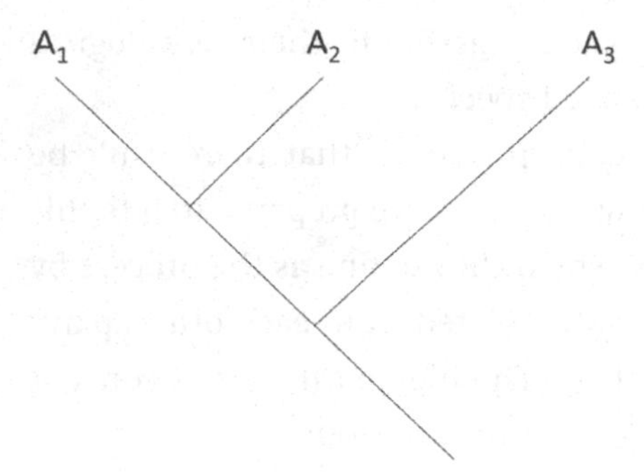

Figure 7.4 The genealogy of three conspecific populations.

individuals in A_1 and A_3. Mayr will say that A_2 is now a new species (for clarity, maybe it should be called "B"), and that $A_1 + A_3$ is now a species as well. However, these two "species" cannot be separate leaves on a phylogenetic tree. $A_1 + A_3$ is like Humans+Gorillas. This point affects biological practice, in that many recognized plant species are known to not be monophyletic (Rieseberg and Brouillet 1994); the brown bear is an example from zoology (Baum and Smith 2013, p. 165).

This thought experiment shows that the biological species concept is the wrong species concept for phylogenetics even though it is alive and well in population genetics. Maybe we need to get used to the fact that the word "species" is ambiguous. The term in evolutionary biology would then be like the term "bank" in ordinary English.

7.8 Race

In 1972, Richard Lewontin published a paper called "The Apportionment of Human Diversity." It was vastly influential, both inside biology and out. Here is Lewontin's conclusion:

> It is clear that our perception of relatively large differences between human races and subgroups, as compared to the variation within these groups, is indeed a biased perception and that, based on randomly chosen genetic differences, human races and populations are remarkably similar to each other, with the largest part by far of human variation being accounted for by the differences between individuals. Human racial classification is of no social value and is positively destructive of social and human relations. Since such racial classification is now seen to be of virtually no genetic or taxonomic significance either, no justification can be offered for its continuance.

Lewontin's critique of the biological reality of race was timely. For many people, Nazi "racial biology" had poisoned the concept of race, as did the racism they saw in their own societies. It was gratifying that a concept with so destructive an impact on human wellbeing could be shown to be a pseudo-scientific construct.

Lewontin's paper grew out of that historical context, but it had a more proximate cause. In 1969, Arthur Jensen published a paper called "How Much Can We Boost IQ and Scholastic Achievement?" arguing that there are racial differences in intelligence, that these differences have a genetic component, and that attempts to reduce those differences are bound to fail. Lewontin (1970a) argued that Jensen's reasoning was deeply flawed. Lewontin's 1972 paper was an attempt to go deeper – to show how one should go about testing whether racial distinctions make biological sense. Lewontin did not discuss intelligence in the 1972 paper.

The title of Lewontin's article refers to *human* diversity, but his topic was more specific – human *genetic* diversity. Lewontin began by laying down four conditions of adequacy that a measure of diversity should satisfy:

> (1) It should be a minimum (conveniently, 0) when there is only a single allele present so that the locus in question shows no variation. (2) For a fixed number of alleles, it should be maximum when all are equal in frequency; this corresponds to our intuitive notion that the diversity is much less, for a given number of alternative kinds, when one of the kinds is very rare. (3) The diversity ought to increase somehow as the number of different alleles in the population increases. Specifically, if all alleles are equally frequent, then a population with ten alleles is obviously more diverse in any ordinary sense than a population with two alleles. (4) The diversity measure ought to be a *convex function* of frequencies of alleles; that is, a collection of individuals made by pooling two populations ought always to be more diverse than the average of their separate diversities, unless the two populations are identical in composition (p. 388).

Given this, Lewontin chose Claude Shannon's concept of information (H) as the appropriate measure. If there are n alleles at a locus in a population with frequencies $p_1, p_2, \ldots, p_n$, then the population's H value for that locus is $-\sum_i p_i(\log(p_i))$, where "log" denotes the base-2 logarithm.[23] If a race is a subgroup of a species, and a population is a subgroup of a

[23] If there are n alleles at a locus in a population, the highest value for H obtains when each has a frequency of $\frac{1}{n}$, and if one population has n alleles each with a frequency of $\frac{1}{n}$ and

race, then $H_{species} \geq H_{race} \geq H_{population}$. The first H is not an average – it's simply the diversity among all the individuals in a single species. The last two H's are averages; H_{race} describes the average diversity *among* individuals in the same race, averaging over races; $H_{population}$ describes the average diversity *among* individuals in the same population, averaging over populations.

Lewontin then describes what *differences* between H values mean. The first difference is

$H_{species} - H_{race}$ = *within*–species diversity–*within*–race diversity

The bigger this difference, the more diversity *among* races contributes to species diversity. The second difference is

$H_{race} - H_{population}$ = *within*–race diversity–*within*–population diversity.

The bigger this difference, the more diversity *among* populations contributes to racial diversity. Lewontin then defines three proportions:

- Within populations = $\frac{H_{population}}{H_{species}}$
- Among the populations in a race = $\frac{H_{race} - H_{population}}{H_{species}}$
- Among the races in a species = $\frac{H_{species} - H_{race}}{H_{species}}$

The sum of these three quantities is 100 percent. Lewontin's big question is: which of the three proportions is biggest?[24]

Lewontin's next step is to decide which human populations he will study. His choice is

> … to count each population included as being of equal value and to include, as much as possible, equal numbers of African peoples, European nationalities, Oceanian populations, Asian peoples, and American Indian tribes. Both of these choices will maximize both the total human diversity

another has m alleles each with a frequency of $\frac{1}{m}$, the former has more diversity precisely when n>m. Shannon's information concept H is often called "Shannon–Weaver entropy." The larger a population's H-value, the more uncertain you are about the characteristics a randomly sampled individual will have, and the more information you gain by looking at that individual.

[24] See Novembre (2022) and Winther (2023) for more details on Lewontin's analysis.

> and the proportion of it that is calculated between populations as opposed to within populations. This bias should be borne in mind when interpreting the results (p. 385).

Lewontin then describes how he will understand the concept of race:

> I have chosen a conservative path and have used mostly the classical racial groupings with a few switches based on obvious total genetic divergence. Thus, the question I am asking is, "How much of human diversity between populations is accounted for by more or less conventional racial classification?" (p. 386)

Lewontin uses what he says are the customary names for those races – Caucasians, Black Africans, Mongoloids, South Asian Aborigines, Amerinds, Oceanians, and Australian Aborigines. He uses data from 169 populations falling into these seven categories. U.S. Blacks were counted as Black Africans, and populations in the Middle East were sometimes counted as Caucasian and sometimes as Black African.

Finally, Lewontin needs to identify the genes whose pattern of diversity he will study. He chooses seventeen loci that affect blood group. Here is Lewontin's conclusion:

> The results are quite remarkable. The mean proportion of the total species diversity that is contained within populations is 85.4% ... Less than 15% of all human genetic diversity is accounted for by differences between human groups! Moreover, the difference between populations within a race accounts for an additional 8.3%, so that only 6.3% is accounted for by racial classification (p. 396).

Lewontin was writing before the birth of molecular biology. Recent molecular biology has provided massive data sets that cover entire individual genomes, and what now counts as genetic variation is much more fine-grained than anything Lewontin could have accessed. However, Lewontin's conclusion about diversity has remained in place. Though recent studies have often avoided using the word "race," the consensus is that human genetic diversity is almost entirely diversity *within* populations; genetic differences *among* populations are very small (see, for example, Rosenberg et al. 2002 and Li et al. 2008).[25]

[25] Edwards (2003) accuses Lewontin (1972) of reasoning fallaciously – of taking his finding about within- and between-race diversity to entail that it is close to impossible

Does Lewontin's finding, and its confirmation by newer and better studies, demolish the idea that races are biologically real? In light of the discussion earlier in this chapter of systematic philosophies, you can see that Lewontin's concept of race is *phenetic*; whether races exist is to be decided by looking at patterns of similarity within and among groups of individuals. What happens to the question of race's biological reality if pheneticism is abandoned? If cladistics is the right approach to taxonomy, we need to see what the concept of race looks like if races are understood genealogically.[26]

Another question about Lewontin's conclusion concerns his "conservative" decision to use "classical" racial categories. Philosophers of psychology often distinguish *folk* psychology from *scientific* psychology. The former describes what people are doing in everyday life when they talk about beliefs, desires, emotions, and sensations; a science of psychology may need to supplement, modify, or replace some or all these familiar categories. Perhaps everyday concepts of race need a similar upgrade. Even if the races that everyday people talk about lack biological reality, perhaps new within-species taxa can pass scientific muster.

Around the time of Lewontin's paper, biologists started using genetic data to construct phylogenetic trees of human populations. Luigi Cavalli-Sforza (1991) describes an early study of this type in which data from people in the seven groups depicted in Figure 7.5 were used to infer a phylogenetic tree. He chose populations for his sample that he thought were mostly reproductively isolated from each other; he included Europeans and Amerinds as distinct groups, but African-Americans were excluded. Notice that African is a monophyletic group in the tree, but Asian is not. Cavalli-Sforza declined to use the term "race" to describe his conclusion.

It is important to bear in mind a feature of phylogenetic inference that applies to all scientific inferences. The inferences you draw are only as good as the observational data you have and the methods you use to

to infer which population or populations an individual's ancestors came from, given enough data about the individual and the populations. Lewontin did not reason in this way; he was not trying to solve the second inference problem. For further discussion, see Feldman and Lewontin (2008).

[26] Indeed, even within the framework of pheneticism, there may be an argument for races having biological reality. Quayshawn Spencer (2014) endorses that thesis based on the phenetic clustering analysis carried out by Rosenberg et al. (2002).

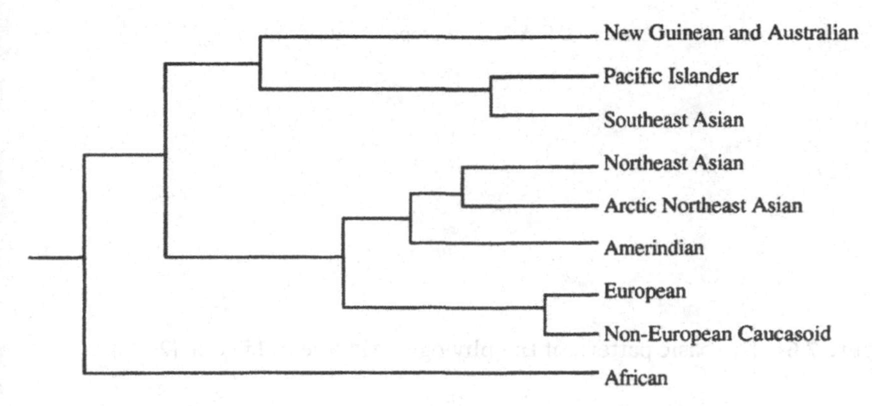

Figure 7.5 Cavalli-Sforza's phylogeny of human groups.

interpret them. Cavalli-Sforza didn't have a large random sample from the nine groups represented on the tips of the tree in Figure 7.5. Rather, he had data on *some* individuals from *some* of the populations in those nine groups, and he was using the genetic data that was available at the time. If you use a data set that is better than the one that Cavalli-Sforza had on hand, or if you use a better inference procedure for interpreting your new data set, your conclusion about what the best tree is might differ dramatically from the one that Cavalli-Sforza identified.

A good example of more and better data on more populations can be found in Li et al. (2008), who used a data set on 938 individuals from 51 populations in the Human Genome Diversity Panel; each individual was characterized by its state at 650,000 common single-nucleotide polymorphism (SNP) loci. That's over 600,000,000 pieces of data! The individuals in each population allowed the authors to estimate the frequencies of traits in that population. Using the assumption that these traits have no effect on an individual's fitness, the authors used a drift model of the evolutionary process to find the phylogenetic tree for the 51 populations that maximizes the probability of the data. When you look at the leaves of this maximum likelihood tree, two striking findings emerge if you classify the 51 sampled populations by using seven broad geographical categories that describe where the people in the sample live – Oceania, America, East Asia, Central/South Asia, Europe, Middle East, and Africa. Rather than reproducing their 51-leaf tree in all its detail, I constructed Figure 7.6, which summarizes the simple pattern in Li et al.'s inferred phylogeny. If you scan the Figure 7.6 tree from lower-right to upper-left, you see the broad sweep of human

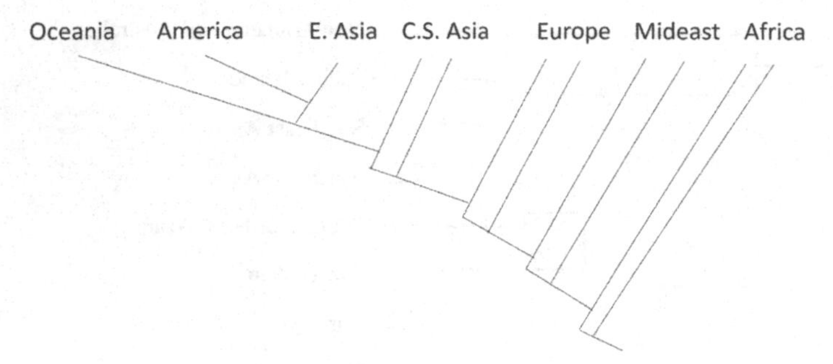

Figure 7.6 The basic pattern of the phylogenetic tree in Li et al. (2008).

migration – African origins, then migration to the Middle East, then to Europe and Asia, and from Asia to the Americas and to Oceania.

The other striking fact about this tree is that only three of the seven geographical labels on the leaves of the tree correspond to monophyletic groups. Oceania, America, and East Asia are monophyletic, but Central/South Asia, Europe, Middle East, and Africa are not. As was true in connection with Cavalli-Sforza's tree, Li et al.'s phylogeny allows you to test whether "conventional races" correspond to taxonomic categories that are biologically real in the sense defined by cladistics.

There is an interesting further step in Li et al.'s analysis. They observed that the sampled individuals in each of the seven geographical locales are more genetically similar to each other than they are to sampled individuals in other locales. From this, they postulated the existence of seven ancestral populations that existed at the same seven locales and that had the same gene frequencies as those found in the sample populations.[27] These inferred ancestral groups are said to have existed in the interior of the tree in Figure 7.6. I'll call these ancestors "recent" to distinguish them from the more ancient ancestor that all the individuals in the sample share; this more ancient common ancestor existed at the root of the tree. All the individuals in Li et al.'s sample have African ancestors, but only some of them have *recent* African ancestors. The authors then used these postulated ancestral populations to say, for each individual in the sample, how much of that individual's genome is drawn from each of those seven postulated ancestral populations. Their conclusion is that some sample individuals

[27] See the discussion in Tang et al. (2005) of "pseudo-ancestors."

have all their genes from just one recent ancestral population, but many have ancestors from more than one.[28]

You may well ask why the seven sample groups with their observed gene frequencies and geographical locations justify the postulation of seven groups of ancestors that had the same gene frequencies and lived in the same locations. After all, groups change trait frequencies and people migrate. Li et al. don't say much to justify this inference, but they could have affirmed the following inequality:

For each of the seven locales,
Pr(the sampled individuals now in locale L have gene frequencies F | there existed an ancestral group that lived in L and had gene frequencies F) >
Pr(the sampled individuals now in locale L have gene frequencies F | there did not exist an ancestral group that lived in L and had gene frequencies F)

The Law of Likelihood describes how this inequality should be interpreted.[29] However, this likelihood argument is not the last word. The postulate of these ancestral populations can be evaluated further by finding data on the genetic characteristics of individuals who once lived in those seven locales. Although DNA degrades with time, there now are tools for recovering and analyzing ancient DNA.

If a genealogy is strictly tree-like, the concept of monophyly can be used to generate a hierarchical classification; the more reticulations there are, the less hierarchical a classification will be if it is based on monophyly. Robin Andreasen (1998) used cladistics to assess the biological reality of race and concluded that human genealogy started off as a bifurcating tree, and then substantial migrations among branches led the genealogy to become highly reticulate. This is why she says that "races once existed, but they are on their way out." But perhaps what has waned in this historical process is the plausibility of a *hierarchical* sub-specific classification. What remains is the existence of *monophyletic* groups; it's just that they partially overlap. The result is that many contemporary individuals belong to multiple disjoint ancestral groups. This possibility is reflected in Li et al.'s analysis if races are defined by an individual's *recent* ancestry.

[28] Li et al.'s (2008) colorful dendrogram makes this point vivid.

[29] This likelihood argument for postulating seven ancestral groups, their geographical location, and their gene frequencies begins with the assumption that there are seven such groups of individuals in the sample. Where did the choice of *seven* come from?

Whether or not races have biological reality, it's possible that races have *psycho-social* reality. Perhaps they exist because individuals identify with each other and also because they do the very opposite. The positive emotions of pride and solidarity contrast with the negative emotions of fear, disrespect, and hatred; these emotions come together to create and sustain psycho-social races. Social institutions and government policies do the same. Lewontin and his successors took the question of whether races exist to be answered by genetics; if psycho-social races exist, they do so because of our minds and actions (James and Burgos 2022).[30] Lewontin, Rose, and Kamin (1990) attacked genetic determinism in a book called *Not in Our Genes*; perhaps this title will have a second life in a book about race.

7.9 Inferring Phylogenies

The cladistic idea that super-specific taxa are monophyletic groups is now the received view in the taxonomy of eukaryotes, but there is another part of cladism that remains controversial. Cladists use a concept of parsimony to infer phylogenetic trees. This is controversial because many biologists use maximum likelihood and other methods to infer phylogenetic relationships and the different approaches sometimes disagree about which tree

Li et al. used a clustering algorithm to form the individuals in their data set into $k = 2$, $k = 3, \ldots, k = 7$ clusters. For each k, the authors did 10,000 computer runs that compared numerous ways of sorting people into k clusters. For each k, these runs agreed about which k-membered cluster had the highest likelihood. However, when they considered clusters for $k = 8$, there was much less consensus among the 10,000 computer runs, and this failure of agreement persisted for $k > 8$. Li et al. set the results for $k > 7$ aside and focused on the maximum likelihood clusters identified for each $k \leq 7$. They then chose to use the $k = 7$ maximum likelihood clustering of individuals to define the grouping of individuals in their sample because its likelihood was greater than the likelihood of each of the $k < 7$ clusters. That's not a very good reason, however, since it was inevitable that the maximum likelihood cluster for $k = 7$ would have a higher likelihood than the maximum likelihood cluster for each $k < 7$. You can understand this by thinking about the extreme case in which you put each individual into its own cluster. There is just one division of this sort and it has a likelihood of unity. Li et al. could have used AIC to compare the maximum likelihood clusters at different values of k. With this procedure, it is not inevitable that the $k = 7$ cluster is better than the ones for $k < 7$.

[30] See Glasgow et al. (2019) for four philosophical perspectives on the biological and psycho-social reality of race. A similar bifurcation may be useful in connection with the contrast between sex and gender (Griffiths 2020, Mikkola 2022).

is best, given the data at hand. The result is that the philosophical task of evaluating competing inference procedures has engaged evolutionary biologists. Here is another example in which a philosophical question is also a question that scientists need to address in order to do their work.

You know from Chapter 4 what maximum likelihood means, but there is a new wrinkle that needs to be recognized. A phylogenetic tree, all by itself, does not tell you what the probability is of the observed characteristics of leaf taxa. For a tree to confer a probability on the observations, auxiliary assumptions are needed about the evolutionary processes at work in branches. Consider, for example, the phylogenetic tree that Li et al. said had a higher likelihood than all the many alternative trees they considered. As mentioned, the authors focused on molecular genetic traits that have no effect on organisms' phenotypes. This led them to use a drift model of the evolutionary process. The drift model M they used contains adjustable parameters. The authors considered numerous conjunctions in which different possible trees (T_1, T_2, ... T_n) were conjoined with M. The parameters in T_1&M, T_2&M, etc. were estimated, so that f(T_1&M), f(T_2&M), etc. could then be compared, where f(T_i&M) is the result of fitting the T_i&M model to data. If f(T_1&M) is the likeliest conjunction, T_1 was then declared the tree of maximum likelihood. The parameters in M are what statisticians call *nuisance parameters*, meaning that Li et al. needed to deal with them, even though their real goal was to evaluate tree topologies. The pattern just described is a probabilistic analog of a familiar idea in philosophy of science that is now called Duhem's thesis, named for Pierre Duhem who claimed in his 1914 book *The Aim and Structure of Physical Theory* that theories in physics do not, by themselves, make observational predictions. What makes a prediction is a conjunction of a physical theory with auxiliary assumptions. Willard Van Orman Quine (1953) generalized Duhem's thesis, saying that it applies to all scientific theories. Duhem and Quine were thinking of deductive connections between hypotheses and observations, whereas the relation of tree hypotheses to observation is probabilistic.[31]

[31] Li et al. were able to deal with Duhem's thesis because they believed that the drift model M they used was correct for the data set they were considering, but what should biologists do if they are uncertain about the processes at work? Here it makes sense to use a model selection approach in which different process models (M_1, M_2, ... M_n)

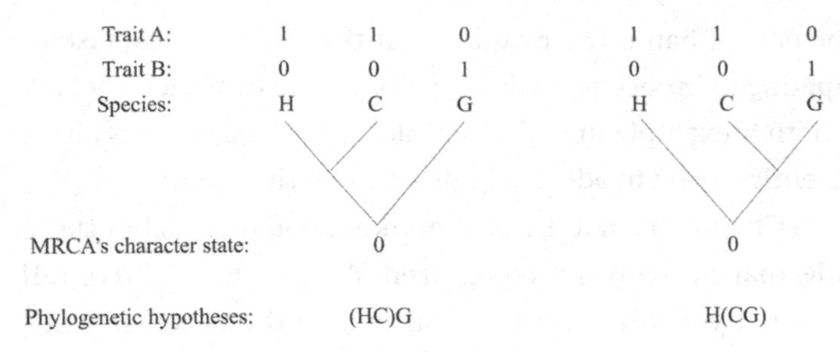

Figure 7.7 Two phylogenetic hypotheses and two traits.

Given this brief description of how likelihood methods are put to work in phylogenetic inference, let's now consider what cladists mean when they say that the best phylogenetic tree is the one that is most parsimonious. As an example, consider the observed characteristics of current humans, chimpanzees, and gorillas. If the genealogy is a bifurcating phylogenetic tree, there are three possibilities: (HC)G, H(CG), and (HG)C. Figure 7.7 depicts two of them and two dichotomous traits. The states for each trait are 0 and 1, and for each trait, the most recent common ancestor of the three species is assumed to be in state 0; 0 is the ancestral state and 1 is the derived state. Trait A involves a *synapomorphy*, meaning that two of the leaf species are in the derived state 1. Trait B involves a *symplesiomorphy*, meaning that two of those species are in the ancestral state 0.

The method of maximum parsimony says that the best phylogenetic hypothesis is the one that requires the fewest changes in the tree's interior to explain the observed character states of the leaves. With respect to Trait A, (HC)G is more parsimonious than H(CG), since the former hypothesis entails that there was at least one change in character state, while the latter entails that there were at least two. Trait B is different; the two hypotheses provide equally parsimonious explanations. The fundamental

are considered, perhaps by using AIC (Section 5.4). This approach has been used by Kishino and Hasegawa (1989), Posada and Crandall (2001), and Posada and Buckley (2004). Instead of accepting the conjunction of a process model and a tree (M_i & T_j) that has the best AIC score and rejecting all the other conjunctions, you can list these conjunctions in order of their AIC scores, from best to worst, and see whether the tree in the best conjunction is also present in the next n conjunctions; the larger n, the more robust your inference is that that tree is best.

idea behind cladistic parsimony is that synapomorphies provide evidence of phylogenetic relationship whereas symplesiomorphies do not.[32]

Cladists often say that synapomorphies are homologies (a concept discussed in Section 7.3), but I think that is a mistake, for two reasons. First, synapomorphies are supposed to be observations that provide evidence about phylogeny; you should be able to see that Humans and Chimpanzees are in the derived state while Gorillas are in the ancestral state of trait A without already knowing that (HC)G is true. Second, even if the (HC)G hypothesis is true and the ancestor was in state 0, the derived state of trait A shared by humans and chimpanzees *might be a homology, but it need not be*; as noted earlier, trait A could have evolved from 0 to 1 in each of the H and C branches independently. In contrast, the H(CG) hypothesis entails that the 1 state in trait A exhibited by humans and chimpanzees *must be a homoplasy.*

Using cladistic parsimony to infer the genealogy of humans, chimpanzees, and gorillas requires that you know, for each of the traits in your data set, which state is ancestral and which is derived. For example, in Figure 7.7, cladistic parsimony says that the 1-1-0 pattern exhibited by trait A favors (HC)G over H(CG) only if 0 is the state of the three species' most recent common ancestor.[33] But how is one to know this? The most important method that cladists use to answer this question is *the method of outgroup comparison*, which is depicted in Figure 7.8. Suppose you don't know what the genealogy is of the three species in the ingroup, but you do know that they are all more closely related to each other than any of them is to the species in the outgroup. Suppose you observe that all these outgroup species are in state 0. The method of outgroup comparison tells you to infer that the most recent common ancestor M of the ingroup was also in state 0. This inference is justified by cladistic parsimony, which leads to the big question: What justifies using cladistic parsimony?

Niles Eldredge and Joel Cracraft (1980) and Ed Wiley (1981) defended the method by arguing that the most parsimonious tree is the one that

[32] Notice that cladistic parsimony is a methodology that applies only to the evaluation of hypothesis about genealogies; this raises the question of whether the justification of cladistic parsimony is the same as the justification for "the principle of parsimony" or "Ockham's razor" when the latter are used to evaluate hypotheses that are not genealogical. I discuss examples of the latter in Sober (2015).

[33] Here "1-1-0" means that H is in state 1, C is in state 1, and G is in state 0 with respect to the trait in question.

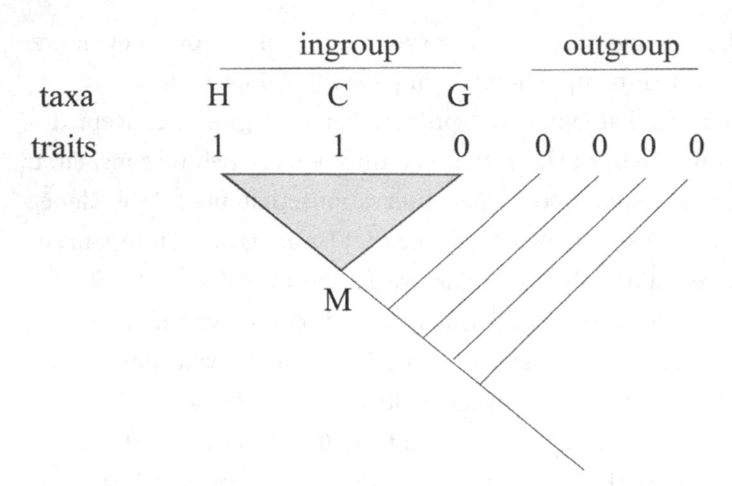

Figure 7.8 The method of outgroup comparison.

is least falsified by the observations, thinking that they were applying Karl Popper's (1959) influential concept of falsifiability. The problem with this suggestion is that a hypothesis that is falsified (in Popper's sense of that term) by even one observation is false, period (Sober 1988c). What cladists probably meant was that the most parsimonious hypothesis is the one that is best supported, but then the question arises of what "support" means here. Is it Bayesian confirmation, or favoring in the sense of the Law of Likelihood, or some other epistemic concept? Steve Farris (1983) took another approach, arguing that the most parsimonious genealogy is the one with the greatest explanatory power. With respect to trait A in Figure 7.7, Farris is saying that (HG)C provides a more powerful explanation of the 1-1-0 observation than H(CG) provides. This suggestion raises the question of what explanatory power means in this context. It is uncontroversial that (HC)G has the following property: if it were true, it would be one part of the explanation of the observed 1-1-0 pattern; the other part would be a description of the processes going on in the branches of that tree. Unfortunately, the same can be said of H(CG).[34]

Even if these proposed defenses of cladistic parsimony fail, maybe there are others that do the job. One possibility to consider is that cladistic

[34] **Question:** Considering just trait A in Figure 7.6, you can see that (HC)G requires only one change, whereas H(CG) requires two. Given the fact that "there was at least one change" is more probable than "there were at least two changes," does that show that (HC)G is more probable than H(CG)?

parsimony has a likelihood justification. Applied to the example of humans, chimpanzees, and gorillas, this would mean that

> Pr(data | (HC)G) > Pr(data | H(CG)) if and only if the number of changes in character state that (HC)G needs to postulate in order to explain the data is less than the number of changes in character state that H(CG) needs to postulate.

Whether this biconditional is true depends on the model of the evolutionary process one considers. As noted, the two tree topologies, taken all by themselves, don't tell you whether the likelihood inequality is true. However, if you use the five assumptions described in Chapter 4 to evaluate common ancestry and separate ancestry hypotheses, and add the further assumption that simultaneous branches obey the same probabilistic rules of evolution, you can deduce that

- Pr(1-1-0 | (HC)G) > Pr(1-1-0 | H(CG)) if 0 is the ancestral state.

This means that the cladist's claim about the evidential meaning of synapomorphies coincides with the likelihoodist's. Unfortunately, that same process assumptions also entail that

- Pr(0-0-1 | (HC)G) > Pr(0-0-1 | H(CG)) if 0 is the ancestral state,

which means that the cladist and the likelihoodist part ways when it comes to assessing the evidential significance of symplesiomorphies.[35] These findings say nothing about how likelihood and cladistic parsimony are related to each other when the assumption about simultaneous branches is replaced.

Joe Felsenstein (1973) proved the following result, which is far more general than what I've just said about single characters and three leaf taxa:

> (F) For any data set D and for any phylogenetic trees T_1 and T_2 where T_1 requires fewer changes in character state to explain D than T_2 requires, if all the branches in T_1 and T_2 have very low probabilities of changing state, then $Pr(D \mid T_1) > Pr(D \mid T_2)$.

If you are a likelihoodist, this may sound like good news for cladistic parsimony, but Felsenstein notes that there is a fly in the ointment. The assumption of low probabilities of change predicts that the maximum parsimony

[35] A derivation of these two results can be found in Sober (2015, pp. 200–202)

trees that biologists have identified for numerous different data sets on different sets of leaf taxa should almost always require very few homoplasies. That is manifestly not the experience that phylogeneticists have. Felsenstein concluded that the assumption of very low probabilities of change on all traits on all branches is false.

Biologists who were parsimony skeptics often took Felsenstein to have shown that parsimony *assumes* that changes on branches are rare, but if you look carefully at the F proposition, you'll see that this interpretation is unwarranted. Felsenstein derived a *sufficient* condition for parsimony to have a likelihood rationale. He did not identify a *necessary* condition.

The fact that Felsenstein identified only a sufficient condition means that there might be a different set of assumptions about the evolutionary process that also would suffice for parsimony to have a likelihood rationale. That possibility remained hypothetical until Chris Tuffley and Mike Steel (1997) described a model of the evolutionary process, very different from Felsenstein's, that also suffices for cladistic parsimony to have a likelihood justification. They called their construct the "no common mechanism" model. It assumes that each trait on each branch has its own probability of changing state and that all changes involve neutral evolution, so changing from state i to state j has the same probability as changing from state j to state i.

The Felsenstein model does not assume neutral evolution but the Tuffley–Steel model does. Felsenstein's model assumes that all traits on all branches have low probabilities of changing state, whereas Tuffley and Steel's model permits probabilities of change to be substantial. These findings have two interesting negative consequences about what cladistic parsimony presupposes. It is *false* that cladistic parsimony assumes that change is improbable, and it is *false* that cladistic parsimony assumes neutral evolution (Sober 2004). In drawing these two conclusions, I am assuming that the Law of Likelihood is correct.

7.10 Inferring the Character States of Ancestors[36]

In addition to using cladistic parsimony to infer phylogenetic trees, biologists often use that methodology to infer the character states of ancestors. For example, if present-day human beings and chimpanzees both

[36] In this section, I use prose from Sober (2015, pp. 170–171).

have trait T, what's the best hypothesis about whether their most recent common ancestor (MRCA) also had trait T? Parsimony sanctions the conclusion that the MRCA has T. Does that conclusion have a likelihood justification?

The answer is *yes*, if you are prepared to view trait evolution as a Markov process. The Markov idea, widely used by statistically minded phylogeneticists, is that lineages can be treated as objects that have probabilities of changing state in very small instants of time. Since these instants are very brief, the chance of a lineage's changing in an instant is very small. If the character in question is dichotomous (with states 0 and 1), there are two instantaneous probabilities of change:

u = Pr(lineage is in state 1 at time t + 1 | lineage is in state 0 at time t)
v = Pr(lineage is in state 0 at time t + 1 | lineage is in state 1 at time t)

Markov models of evolution compute the probability that a lineage will end in a given state j, given that it starts in state i and there are t units of time between start and finish. There are four such probabilities for a dichotomous character:

$$\Pr_t(\text{end in state } 1 \mid \text{start in state } 0) = \frac{u}{u+v} - \frac{u}{u+v}(1-u-v)^t$$

$$\Pr_t(\text{end in state } 0 \mid \text{start in state } 0) = \frac{v}{u+v} + \frac{u}{u+v}(1-u-v)^t$$

$$\Pr_t(\text{end in state } 0 \mid \text{start in state } 1) = \frac{v}{u+v} - \frac{v}{u+v}(1-u-v)^t$$

$$\Pr_t(\text{end in state } 1 \mid \text{start in state } 1) = \frac{u}{u+v} + \frac{v}{u+v}(1-u-v)^t$$

These four probabilities are *branch* transition probabilities. The first two probabilities sum to one, and so do the third and fourth. In each equation, the first addend fails to mention the amount of time t between the lineage's start and finish. When time is short, the values of these transition probabilities are mainly determined by the lineage's initial state; in that case the lineage has a very high probability of ending in the state in which it began. Notice that when $t = 0$, the first and third probabilities displayed above have values of 0 and the second and fourth have values of 1. As the duration of the lineage increases, the second addend shrinks towards zero as t increases, since $(1 - u - v) < 1$. This means that the process plays a progressively larger role in influencing the probability of the final state and the initial condition of the lineage is steadily "forgotten". In the limit

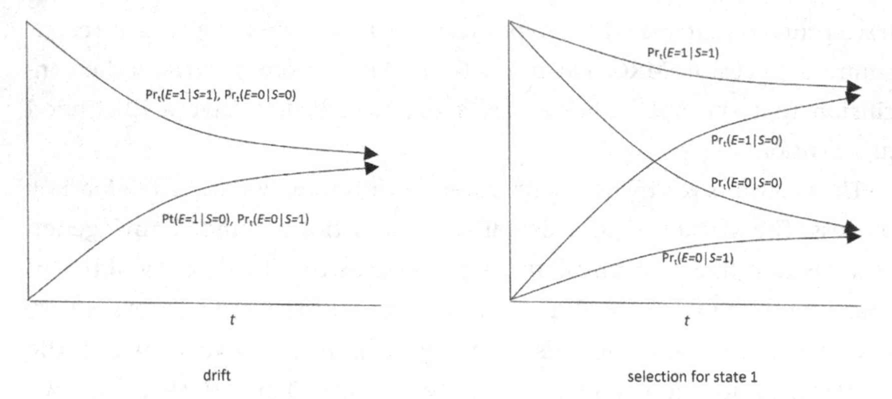

Figure 7.9 Markov models for drift and selection in a lineage that starts in state S and ends in state E.

as t approaches infinity, the probability of ending in state 1 is the same regardless of the lineage's starting state; ditto for the probability of ending in state 0.[37]

Each of the four conditional probabilities described takes account of all the possible flip-flops that might occur in the lineage between start and finish. For example, the second and fourth probabilities take account of the possibility that there were 0 changes, 2 changes, 4, 6, and so on. The four equations entail the following inequality, which I discussed in Section 4.4:

(COR) $\Pr_t$(descendant is in state i | ancestor is in state i) >
$\Pr_t$(descendant is in state i | ancestor is in state j).

Ancestor and descendant states are *positively correlated*.

This Markov model can be used to describe the difference between drift and selection. Drift means that $u = v$; selection for state 1 means that $u > v$. The impact of that difference on branch transition probabilities is depicted in Figure 7.9. Although COR is true whether there is drift or selection, the following inequality is true when there is drift, but it can be false when there is selection for state j:

Pr(descendant is in state i | ancestor is in state i) >
Pr(descendant is in state j | ancestor is in state i)

[37] Markov *processes* have the Markov *property*; if a lineage evolves from state D in the distant past to state R in the recent past to state P in the present, R screens-off D from P. Screening-off was discussed in Section 4.4.

If the ancestor starts in state i and there is enough time between ancestor and descendant, selection for state j will make change more probable than stasis.

Returning now to the problem of inferring the character state of X and Y's most recent common ancestor, where X and Y are in the same state of a dichotomous character, I hope you can see from COR that the most parsimonious hypothesis about the common ancestor's state is also the hypothesis of maximum likelihood. Maximum parsimony and maximum likelihood are on the same page.

Now consider a different problem. Suppose that descendants X and Y are in different states; X is in state 1 and Y is in state 0. The method of maximum parsimony says that the hypothesis that the MRCA of X and Y was in state 1 and the hypothesis that the MRCA was in state 0 are equally good. To see what the Law of Likelihood says, let's use the Markov model and assume that

$$\Pr(X=1 \mid MRCA=1)=\Pr(Y=1 \mid MRCA=1)=a$$
$$\Pr(X=1 \mid MRCA=0)=\Pr(Y=1 \mid MRCA=0)=b,$$

Since changes on the two branches are independent, you can see that

$$\Pr(X=1 \,\&\, Y=0 \mid MRCA=1)=a(1-a)$$
$$\Pr(X=1 \,\&\, Y=0 \mid MRCA=0)=b(1-b)$$

Notice that $a(1-a)=b(1-b)$ if and only if $(a-b)=(a-b)(a+b)$. Given that $a>b$ (the COR inequality), $(a-b)=(a-b)(a+b)$ precisely when $(a+b)=1$. This last equality says that $a=(1-b)$, which means that $\Pr(X=1 \mid MRCA=1)=\Pr(X=0 \mid MRCA=0)$, so drift, not selection, is the process at work. Thus, if drift is the process, likelihood and parsimony agree that the data do not discriminate between the hypothesis that MRCA = 1 and the hypothesis that MRCA = 0.[38]

What if there is selection for state 1? According to Figure 7.9, selection for state 1 entails that the ordering of the four probabilities depends on the amount of time separating ancestor and descendant:

- When time is short, $a>(1-b)>b>(1-a)$
- When time is long, $a>b>(1-b)>(1-a)$

[38] **Exercise:** Suppose X = 1 and Y = 0 and that a drift process governs the evolution of those trait values in the lineages stemming from X and Y's MRCA, but that X is an extant organism and Y is an ancient fossil. Do parsimony and likelihood agree about what the best estimate is of the MRCA's character state? Explain your answer.

These orderings agree that a is the biggest and (1 – a) is the smallest, with b and (1 – b) falling in between. This means that $a(1 - a) < b(1 - b)$ regardless of whether time is short or long. Thus, when there is selection for trait 1, the hypothesis that MRCA = 1 has a lower likelihood than the hypothesis that MRCA = 0 even though they are equally parsimonious. Now likelihood and parsimony disagree.[39]

7.11 Why Use Likelihood to Evaluate Cladistic Parsimony?

In the previous two sections, I used the Law of Likelihood to evaluate cladistic parsimony. The result was a mixture of good news and bad. The good news is that when simultaneous branches have the same transition probabilities, parsimony is right in the way it interprets synapomorphies and it also is right that a dichotomous trait shared among descendants is evidence that their most recent common ancestor also had that trait. The bad news is that when simultaneous branches have the same transition probabilities, parsimony is wrong in the way it interprets symplesiomorphies and it also is wrong in what it says about an ancestor's character state if descendants differ in their character states and selection is occurring. Friends of the Law of Likelihood will find these results meaningful, but friends of cladistic parsimony may reject the idea that the Law of Likelihood gets to arbitrate whether cladistic parsimony is right or wrong in what it says. Instead of evaluating parsimony by using the Law of Likelihood, they may suggest that the Law of Likelihood should be evaluated by using parsimony as the gold standard.

Cladists have tried to justify parsimony in terms of other, more general, considerations. I mentioned their appeals to Popperian falsification and to explanatory power in Section 7.9. However, maybe cladists should maintain the cladistic parsimony is a primitive postulate, meaning that it can't be justified in terms of anything more fundamental; it is rock bottom. Hennig embraces something like this position when he says that

> the presence of apomorphous characters in different species "is always reason for suspecting kinship [i.e., that the species belong to a monophyletic group], and that their origin by convergence should not be

[39] For further discussion of parsimony versus likelihood in ancestral character reconstruction, see Cunningham et al. (1998).

> assumed a priori".... This was based on the conviction that "phylogenetic systematics would lose all the ground on which it stands" if the presence of apomorphous characters in different species were considered first of all as convergences (or parallelisms), with proof to the contrary required in each case. Rather, the burden of proof must be placed on the contention that "in individual cases the possession of common apomorphous characters may be based only on convergence (or parallelism)" (Hennig 1950/1966, pp. 121–122; square-bracketed material and quotation marks are his).

This should remind you of the discussion of default reasoning in Section 5.4.

A parallel question can be posed about the Law of Likelihood – is it a primitive postulate? I noted in Section 4.2 that Bayesians can justify the law by appeal to the odds formulation of Bayes's theorem. Fisher wasn't a Bayesian, so he would find that argument unconvincing. Rather, Fisher (1938) says that maximum likelihood is a "primitive postulate," though he later came to think that it needs to be judged in terms of a more fundamental consideration that I'll discuss in the next section.

Setting aside this question about the status of the Law of Likelihood, I think there is something odd about the claim that cladistic parsimony is a primitive postulate. I say this for two reasons. First, cladistic parsimony is an inference procedure that is specific to the task of inferring genealogical trees. If cladistic parsimony is a primitive postulate in that context, are there thousands of other such inference principles, each limited in scope to a specific problem in a specific science?[40] That strikes me as implausible. Second, recall the discussion in Chapter 4 of evidence for common ancestry. When two or more species have a common *ancestor*, that is an instance of the more general situation in which two or more events have a common *cause*. This suggests that a good epistemology for inferring common ancestry should also apply to the more general problem of inferring common causes. The question of whether human beings are more closely related to chimpanzees

[40] Philosophers have sometime doubted the existence of "the" scientific method by pointing out that evaluating the impact of observations on competing hypotheses requires background assumptions that are subject-matter specific. I don't think this is a good argument. Bayesianism, for example, recognizes that whether a given observation confirms a given hypothesis almost always depends on subject-matter specific background assumptions, and yet Bayesianism is a *general* framework for interpreting evidence; it is not subject-matter specific. Parallel points apply to the Law of Likelihood and to frequentist statistics. **Question:** Is the fact that inference practices have changed in science a good reason to doubt the existence of "the" scientific method?

than they are to gorillas is one step away from the question of whether these three species have a common ancestor. The latter question should be anchored to a more general epistemology, so the former should be as well.

7.12 Statistical Consistency

Whether or not the Law of Likelihood should be used to evaluate cladistic parsimony, it would be nice if there were an independent criterion that could be used to evaluate them both. Joe Felsenstein (1978) thought he had found one when he asked whether parsimony and likelihood are each *statistically consistent*. The rough idea here is that a consistent estimator will converge on the true value of the quantity being estimated as more and more data are considered. In assuming that an acceptable estimator must be consistent, Felsenstein concurred with Fisher's (1950, p. 11) remark that inconsistent estimators are "outside the pale of decent usage." Fisher (1956, p. 141) reiterates this point, calling consistency "the fundamental criterion of estimation."

Felsenstein's paper was mostly about how parsimony applies to the simple hypothetical example depicted in Figure 7.10. Suppose the true phylogenetic relationship of taxa A, B, and C is that A and B are more closely related to each other than either is to C. You are going to look at numerous dichotomous traits (whose two states are 0 and 1) that A, B, and C have and you know that the most recent common ancestor of the three taxa was in state 0 for all its traits. You then check how often the traits of A, B, and C have the 1-1-0 pattern and how often they are in the 1-0-1 configuration. Cladistic parsimony says that traits of the first type favor (AB)C over (AC)B whereas traits of the second type have the opposite evidential significance. Parsimony will be a statistically inconsistent estimator of the tree topology if

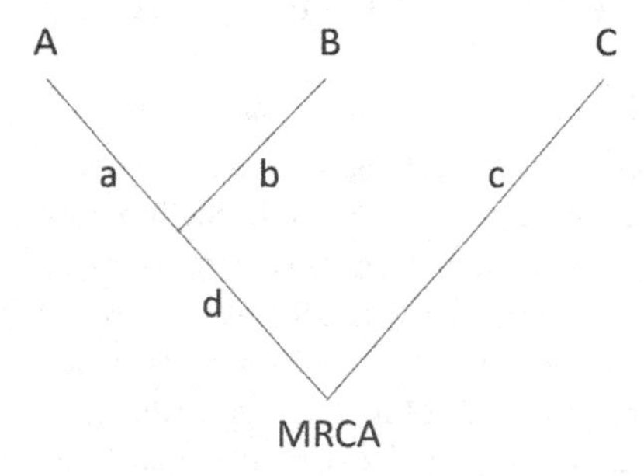

Figure 7.10 Felsenstein's (1978) example.

(INC) Pr(1-0-1 | (AB)C) > Pr(1-1-0 | (AB)C).

This inequality guarantees that as you look at more and more data, the probability approaches one that the traits you examine will have the 1-0-1 pattern more often than the 1-1-0 pattern. This means that cladistic parsimony becomes more and more certain to says that (AC)B is the true phylogeny when, in fact, it is (AB)C that is true. The INC inequality says that parsimony is statistically inconsistent.

Felsenstein's next step is to describe a simple model that entails INC. The model says that all the traits that evolve on a given branch are governed by the same rules of evolution, that evolving from state 1 to state 0 is impossible, and that branches a and c have the same probability of evolving from 0 to 1 and that branches b and d likewise have the same probability of evolving from 0 to 1:

$$\Pr_a(\text{end in state } 1 \mid \text{start in state } 0) = \text{pr}_c(\text{end in state } 1 \mid \text{start in state } 0) = p$$

$$\Pr_b(\text{end in state } 1 \mid \text{start in state } 0) = \text{pr}_d(\text{end in state } 1 \mid \text{start in state } 0) = q$$

A bit of algebra then reveals that INC is true in a region of parameter space in Figure 7.11. In this example, branches a and c are said to exhibit "long branch attraction," meaning that their high probabilities of changing state (as compared with those for branches b and d) lead parsimony to mistakenly conclude that A and C are closely related.

Farris (1983) argued that Felsenstein's result is not a good criticism of parsimony because the model of the evolutionary process that Felsenstein uses is extremely unrealistic. Felsenstein (1978) agrees that his model is unrealistic, but says that it would be naïve to assume that parsimony will be consistent in more realistic models of the evolutionary process, given that it fails to be consistent in his example.[41]

If statistical inconsistency is bad news for parsimony, is the news for likelihood any better? Felsenstein's answer is that "it can be shown quite generally that the maximum likelihood estimation procedure has the property of consistency" (Felsenstein 1978, p. 408). To evaluate

[41] Here's an unsurprising result that is good news for parsimony: if a trait's rules of evolution are the same across simultaneous branches and evolution is Markovian, then Pr[1-1-0 | (AB)C] > Pr[1-0-1 | (AB)C], Pr[0-1-1 | (AB)C] for each character (Sober 2015, pp. 200–202).

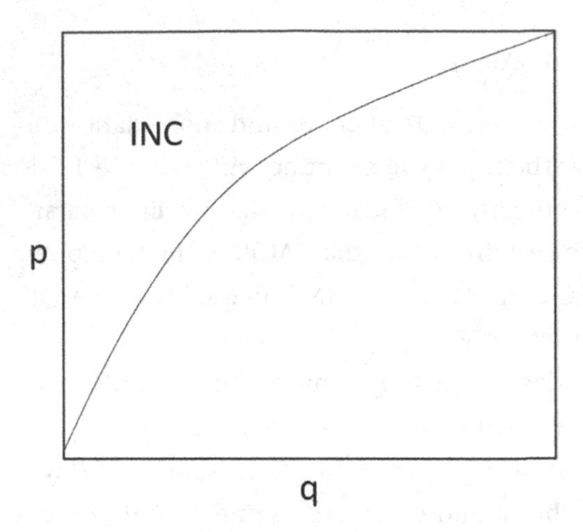

Figure 7.11 Cladistic parsimony will be statistically inconsistent in the INC region as an estimator of the topology shown in Figure 7.10, given Felsenstein's (1978) process assumptions.

whether Felsenstein is right, a more careful characterization of statistical consistency is needed, which I'll provide by returning to coin tossing. What is true is that if tossing the coin obeys an i.i.d. model (meaning that tosses are independent of each other and each toss has the same probability p of landing heads), then the maximum likelihood estimates of p will converge on the true value of p as the coin is tossed repeatedly. That is,

> $ML_{i.i.d.}(p)$ is a consistent estimator of the coin's probability p of landing heads precisely when, for any $\varepsilon > 0$, Pr[s is within ε of p | $MLE_{i.i.d.}(p \mid O_1 \;\&\; O_2 \;\&\; \ldots \;\&\; O_n) = s$ & the i.i.d. model is true & p is the true probability] approaches 1 as n approaches ∞.

Here s is the maximum likelihood estimate of the coin's probability of landing heads obtained by observing the outcomes of n tosses and ε (epsilon) is any margin of error you please. Notice that this definition does not define what it means for ML to be consistent *full stop*; what it defines is the consistency of ML *in the context of the assumed i.i.d. model.*

This definition of a consistent estimator carries over to phylogenetic inference as follows:

> $ML_M(t)$ is a consistent estimator of the tree topology t of a set of taxa precisely when
> $Pr[s = t \mid MLE_M(t \mid O_1 \;\&\; O_2 \;\&\; \ldots \;\&\; O_n) = s$ & M is true & t is the true topology] approaches 1 as n approaches ∞.

Here M is a probabilistic model of how traits evolve on branches and each O_i describes the observed character states of leaf taxa for a trait. The important point about this definition is this: if you estimate the topology by assuming that *M* is true, and *M* is false, then you have no assurance that your maximum likelihood procedure $ML_M(t)$ will converge on the true topology *t* as you gather more and more data (Gaut and Lewis 1995; Steel and Penny 2000). In this case, statisticians say that the model has been *mis-specified.* A similar conclusion follows for coin-tossing. If you assume that the coin is an i.i.d. system and it is not, there is no assurance that $ML_{i.i.d.}(p)$ will converge on the true value of *p*, since then there is no such thing.

Given that statistical consistency is an asymptotic property, the question of whether a method of inference is consistent is different from the question of how often a method concludes from a finite data set that one tree is better than another, when, in fact, the one tree is false and the other is true. Biologists, lacking time machines, have no direct access to the true trees out there in nature, but they attempt to cut this Gordian knot by performing experiments. They use a population of rapidly reproducing microorganisms in their laboratories to found multiple descendant populations, which in turn are used to found descendant populations after that. At each stage, the traits of the different populations can be observed. At the end of this multistep process there are n leaf populations whose phylogenies are known. Different methods of inference are then used to infer a phylogenetic tree from data collected on leaf taxa, and their performance can be judged by seeing how well they do in reconstructing the true phylogeny (e.g., see Hillis et al. 1993). It is important to ask how representative these experiments are of what happens in nature, but numerous such experiments on organisms from different species and in different environments can provide a fuller picture. Hacking (1965, p. 185) wondered whether consistency is a "magical property" of an estimator rather than being a "criterion of excellence." Comparing Felsenstein's (1978) argument with these experiments might help answer Hacking's question.[42,43]

[42] In Sober (1988b), I provide another perspective on this question by constructing a simple example in which a good likelihood inference is statistically inconsistent.

[43] See Haber and Velasco (2022) for further discussion of philosophical questions concerning phylogenetic inference.

8 Adaptationism

8.1 Darwin's Formulation

To analyze Darwin's (1859, p. 6) idea that "natural selection has been the main but not the exclusive cause" of evolution, let's begin with a three-way distinction concerning the impact that selection might have on the evolution of a trait:

(A) Selection was *a* cause of the evolution of trait T in lineage L.

(P) Selection was a *powerful* cause of the evolution of trait T in lineage L.

(O) Selection was the *only* powerful cause of the evolution of trait T in lineage L.

These propositions are listed from logically weaker to logically stronger; P entails A but not conversely, and O entails P but not conversely. Darwin's formulation, when applied to trait T in lineage L, is not on this list; it should be wedged between P and O.[1]

To show that natural selection is the most powerful cause of the evolution of a trait, you need to show that it is more powerful than the other causes in play. When two causes exert opposite influences on an observed outcome, it often is easy to see what it means for one cause to be more powerful than the other. If Sam pushes the billiard ball east while Aaron pushes it west (and there are no other pushers), the motion of the ball will tell you which push is stronger. In Section 5.8. I mentioned the evolution of altruism and

[1] An alternative interpretation of what Darwin meant by "main" is that he meant that selection is the most pervasive cause of evolution, meaning that selection influences the evolution of more traits than any nonselective cause manages to affect. I doubt that this is what Darwin had in mind. Darwin recognized that mutation (the production of new variants) is even more pervasive, in that mutation can produce neutral variation as well as variation in fitness.

selfishness in a meta-population, where trait frequencies vary within and among groups. In this case, group and individual selection both occur, and they push in opposite directions. If altruism increases in frequency in this process, that's evidence that group selection was the more powerful cause.[2]

When this opposing-forces picture fails to apply, it can be less straightforward to say that one cause of an event was more powerful than another.[3] Consider two procedures that a pair of workers might use to build a wall.[4] In the first procedure, one worker brings 60 percent of the bricks to the construction site and the second brings 40 percent, after which each worker mortars the bricks she brought and puts them in place. In the second procedure, one worker brings the bricks and mortar and the other does all the mortaring and placement. It is clear that one worker contributed more than the other in the first scenario, but no such judgment can be made about the second. The workers in the second are doing qualitatively different jobs, each necessary but insufficient for getting the wall built. Neither contributed more than the other. Does this second scenario have an analog in the evolutionary process?

The answer is *yes*, as Darwin was well aware. Three years after the *Origin* appeared, Darwin published *On the Various Contrivances by which British and Foreign Orchids are Fertilised by Insects*. Here's how Darwin describes one of the book's central themes:

> In my examination of Orchids, hardly any fact has struck me so much as the endless diversities of structure – the prodigality of resources – for gaining the very same end, namely, the fertilization of one flower by pollen from another plant. (Darwin 1862/1867, p. 284)

Orchids are hermaphrodites, but they have a variety of adaptations that hinder self-fertilization and help a flower send its pollen to other flowers. Darwin thought that these species often have different adaptations simply because they differed in the variations that happened to be available for natural

[2] If each force exerts a probabilistic influence on change in trait frequencies, you can't *deduce* that altruism must increase in frequency from the hypothesis that group selection is the stronger force. **Exercise:** Use the law of likelihood to describe how an observed change in trait frequency favors one hypothesis about the strengths of opposing forces over another.

[3] The question of what it means for selection to dominate drift is an instance of this problem (Section 5.9).

[4] This example is a modification of an example described by Lewontin (1974, p. 521).

selection to act upon. To illustrate this idea, Darwin describes the relationship of two parts of orchid morphology – the ovarium and the labellum:

> In many Orchids the ovarium (but sometimes the foot-stalk) becomes for a period twisted, causing the labellum to assume the position of a lower petal, so that insects can easily visit the flower; but … it might be advantageous to the plant that the labellum should resume its normal position on the upper side of the flower, as is actually the case with *Malaxis paludosa*, and some species of Catasetum, &c. This change, it is obvious, might be simply effected by the continued selection of varieties which had their ovaria less and less twisted, but if the plant only afforded variations with the ovarium more twisted, the same end could be attained by the selection of such variations, until the flower was turned completely on its axis. This seems to have actually occurred with *Malaxis paludosa*, for the labellum has acquired its present upward position by the ovarium being twisted twice as much as is usual (Darwin 1862/1867, pp. 284–285).

Darwin believed that all current orchids trace back to a common ancestor in which the labellum was above the ovarium. Most of the lineages deriving from that ancestor evolved away from that state and now have the labellum below; it serves as a platform on which insect pollinators can land. To achieve this change, these orchids twisted their stems by 180°. However, a few orchid species subsequently returned to the ancestral arrangement. Some of these revisionists went back the way they came, with the result that their stems have zero twist, but others increased their twist with the result that their stems now twist a full 360°; *Malaxis paludosa* is an example of the latter. I've reproduced Darwin's picture of that orchid in Figure 8.1.

Darwin's explanation of the difference between these two ways of returning to the ancestral arrangement is *not* that there was a selective advantage for one group to untwist, while the other found an advantage in twisting some more. Darwin's idea was that untwisting mutations occurred by chance in some of these groups while further-twisting mutations occurred by chance in others. The difference between orchids with untwisted stems and orchids with 360° twists is due to mutational differences, not to natural selection.[5,6]

[5] My account of Darwin's reasoning in this example is drawn from Beatty's (2006) insightful essay.

[6] A modern genetic version of the same point is that selection for lactose prolongation (the ability to digest lactose in adulthood, not just in infancy) in populations of human dairy farmers resulted in different genes evolving in different groups in Africa, the Middle East, and Europe (Gerbault et al. 2011).

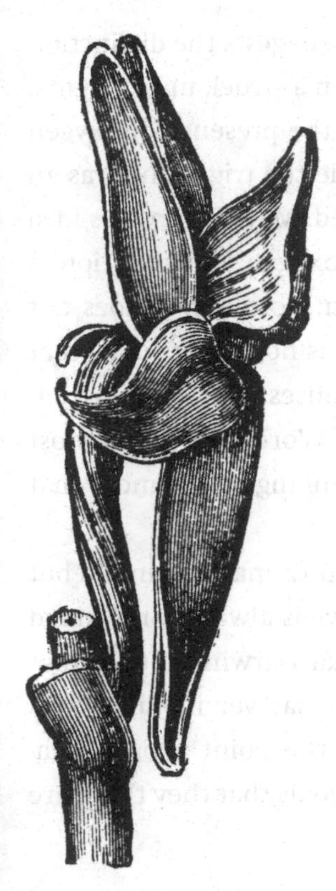

Figure 8.1 The orchid *Malaxis paludosa* has a 360° twist in its stem (Darwin 1862/1877, p. 130).

The evolution of the 360° twist in *Malaxis paludosa* was caused by both mutation and selection. Which was the more important cause? Arguably, if the mutations hadn't occurred, the 360° twisting would not have evolved, and if there had been no selection for getting the labellum to be above the ovarium, the twisting would not have evolved. Darwin's phrase "selection has been the main but not the exclusive cause" does not do justice to what he thought about orchids. The stumbling block in Darwin's formulation is the definite article. What Darwin thought, and what he should have said consistently, is proposition P – that natural selection is *a powerful* cause.

A hint about why Darwin used "the" can be found in a letter he wrote to Joseph Hooker (dated May 29, 1860) in which he tries to clarify

> the part which I believe Selection has played. I have been reworking my pigeons & other domestic animals & I am sure that anyone is right in saying that Selection is <u>the</u> efficient cause, though as you truly say variation is the base of all.

The double underlining is Darwin's.[7] His comment suggests the distinction between *enabling* causes and *triggering* causes.[8] When a struck match lights, the striking might be said to be the trigger while the presence of oxygen is enabling. The oxygen was present for ages, while the triggering was an event that happened right before the match ignited. Maybe Darwin's idea is that standing mutational variation is like the oxygen and selection is like the striking. I don't object to this suggestion; however, it does not undercut my point that both are causes, and there is no reason to say that triggering causes are more potent than enabling causes. The assassination of Archduke Ferdinand was the triggering cause of World War 1, but most historians think it was less important than the entangled alliances that were present before 1914.

The details of the orchid example aren't needed to make a simple but important point. What natural selection can achieve is always constrained by the available variation. I noted in Chapter 5 that Darwin thought that plant and animal breeders were able to achieve whatever modifications they set their sights on, but this doesn't negate the point about mutational variation setting limits. Breeders formulate goals that they think are achievable; they work with the variations they see.

8.2 What Is Adaptationism?

The three propositions A (= *a* cause), P (= a *powerful* cause), and O (= the *only* powerful cause) are about a single trait T in a single lineage L, but they each can be generalized in various ways. The logically strongest generalization of O says that O is true of all the traits that organisms now have. A more modest generalization would restrict O to all phenotypic traits. This phenotypic version of O can be made even more modest by replacing "all" with "almost all," and that can be weakened further by moving from "almost all" to "most" and then to "many." Ditto for propositions A and P.

[7] Harvey had been corresponding with Hooker about the theological implications of Darwin's theory. Hooker sent Harvey's letter to Darwin. Darwin's reply can be found at www.darwinproject.ac.uk/letter/?docId=letters/DCP-LETT-2816.xml&query=huxley. I thank Hayley Clatterbuck for drawing my attention to this letter.

[8] Dretske (1988) distinguishes "structuring" from "triggering" causes, though the details of how he understands structuring causes are not what I want to use here.

I think the thesis of adaptationism should be understood as asserting a fairly strong form of O. It is restricted to phenotypes, but it declines to make a claim about all of them:

> **Ontological adaptationism:** Almost all the phenotypes that organisms now have evolved with natural selection as the only important cause.

I use the term "ontological" because this *ism* makes a claim about what there is. I have not used "most" here because I think adaptationists generally want to postulate something more than a bare majority. I formulate adaptationism as a generalization of O, not of P, since critics of adaptationism often grant that P is right as a claim about phenotypes. For example, Stephen Jay Gould and Richard Lewontin (1979), in their influential paper "The Spandrels of San Marco – A Critique of the Adaptationist Programme," say they agree with Darwin that selection has been the main but not the exclusive cause of evolution.

What does "evolved" encompass in this formulation of ontological adaptationism? Are we talking just about the changes in trait frequency that occur until a mutation reaches fixation, or do we want to also include the fact that the trait remains at high frequency afterwards? I think adaptationists typically have both of these in mind. If so, a trait that reached fixation just by drift but then was retained at 100 percent by stabilizing selection would not conform to the adaptationist picture.

In earlier chapters, I've used the rule of thumb that fitter phenotypes increase in frequency and less fit traits decline if selection is in control of the evolutionary process. Adaptationists and anti-adaptationists can agree that this thesis is correct; their agreement is possible because of the word "if." The two camps disagree about how often the antecedent (the "if" part) of the conditional is satisfied. The previous section described one way in which selection can fail to be in total control. I'll describe other sources of failure in a moment.

Ontological adaptationism is not the same as the thesis that almost all phenotypes are adaptations. Recall from Section 2.7 that "trait T is an adaptation" just means that T evolved because there was selection for that trait. This does not entail that selection was the only powerful cause of the trait's evolution. However, if selection for trait T was the only important cause of the evolution of that trait, then T is an adaptation. Thus ontological adaptationism entails that adaptations are abundant, but the converse entailment does not hold.

Ontological adaptationism has a methodological implication. It tells you that a certain sort of modelling practice will be fruitful. For example, in thinking about why zebras run fast rather than slow, suppose you ignore how genetics affects an organism's running speed – you assume that zebras reproduce asexually, with fast parents always having fast offspring and slow parents always having slow, and you assume that zebras live in enormous herds. The result is a model that is both incomplete (owing to the absence of genetic details) and false (owing to the idealizations concerning asexual reproduction and the assumed population size). If you are an adaptationist, you will regard the falsehood and incompleteness of your model as *harmless*. Correcting them wouldn't much affect the predictions of your model (Sober 1993). Your model resembles the model of a ball rolling down an incline plane where the model says that the plane is frictionless; if the plane is close to frictionless, your idealization is harmless. Generalizing from this example yields the following:

> **Methodological adaptationism:** For almost all phenotypic traits, evolutionary models that are incomplete (because they ignore genetics) and that make idealizing assumptions about evolution (e.g., that the organisms described are asexual, that offspring always have the phenotypes of their parents, and that the organisms live in huge populations) can make accurate predictions about which trait values will evolve.

This modelling strategy is what biologists sometimes call *the phenotypic gambit* (Grafen 1991; Smith and Winterhalder 1992). Note that methodological adaptationism goes beyond the anodyne suggestion that in understanding the evolution of phenotypic traits, it's important to formulate and test hypotheses that describe how natural selection influenced their evolution.[9]

[9] Godfrey-Smith (2001) distinguishes three kinds of adaptationism. He calls the first two "empirical adaptationism" and "methodological adaptationism." Combining his two isn't much different from my two combined. Godfrey-Smith's third adaptationism is something new:

> Explanatory Adaptationism: The apparent design of organisms, and the relations of adaptedness between organisms and their environments, are the *big questions*, the amazing facts in biology. Explaining these phenomena is the core intellectual mission of evolutionary theory. Natural selection is the key to solving these problems; selection is the *big answer*. Because it answers the biggest questions, selection has unique explanatory importance among evolutionary factors [italics his].

There is more to be said about adaptationist methodology; some of it concerns the practices of adaptation*ists*, not the commitments of adaptation*ism*, ontological or methodological. In addition to disagreeing with ontological adaptationism, Gould and Lewontin (1979) claim that adaptationists often endorse adaptationist explanations uncritically, in part because they fail to consider alternative hypotheses. Notice that these criticisms don't refute the ontological adaptationism defined above, nor do they undercut methodological adaptationism as I have defined it. Methodological adaptationists can and should subject the models they construct to rigorous testing.

I've said that methodological adaptationism makes sense if ontological adaptationism is true. Note the *if*. In Section 8.5, I'll argue that critics of the ontological thesis also have a good reason to construct and test models that conform to methodological adaptationism's recommendation.

8.3 How Genetics Can Get in the Way

The simplest picture of how genotypes and phenotypes are related in a selection process is that different phenotypes are coded by different genotypes in such a way that natural selection will cause the fittest of the phenotypes to evolve to 100 percent. For example, suppose there are three running speeds that zebras might have (slow, medium, and fast), and the fitness ordering is that fast is fitter than medium which in turn is fitter than slow. Suppose organisms with the gg genotype run slow, those with fg run medium, and those with ff run fast, as shown in Figure 8.2. Suppose further that the three genotypic fitnesses are frequency independent. From these three flat lines you can extract the allelic fitnesses – the fitnesses of f and g – which are depicted by broken lines. The allelic fitnesses are frequency dependent. How did that happen? Notice that when f is rare it is found almost entirely in fg heterozygotes, but when f is common it is found almost entirely in ff homozygotes.[10] In this model, f is fitter than g regardless of the allelic frequencies, and so our familiar rule of thumb says

I suppose that biologists often think this way, but I think it is a matter of taste whether adaptedness is *the* big question. There are sensible alternatives. Consider the suggestion that "the big question" is phenotypic and genetic variation within and among species; or that there are several "big questions" with none of them more important than the other. I agree with Godfrey-Smith's skepticism about this third thesis.

[10] This reprises the point from Section 2.6 that trait fitnesses are averages.

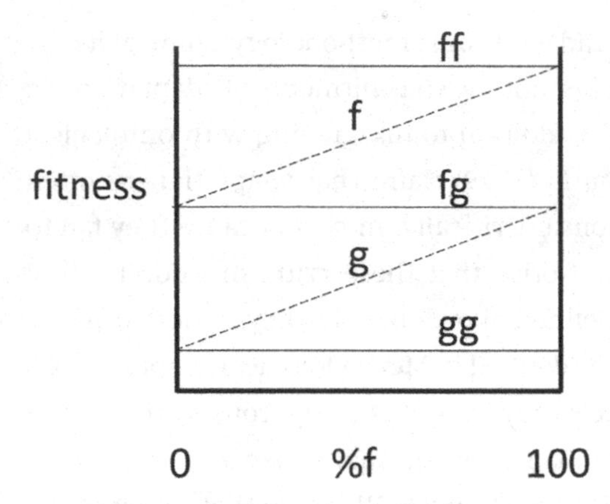

Figure 8.2 A one-locus, two-allele model in which the heterozygote is intermediate in fitness.

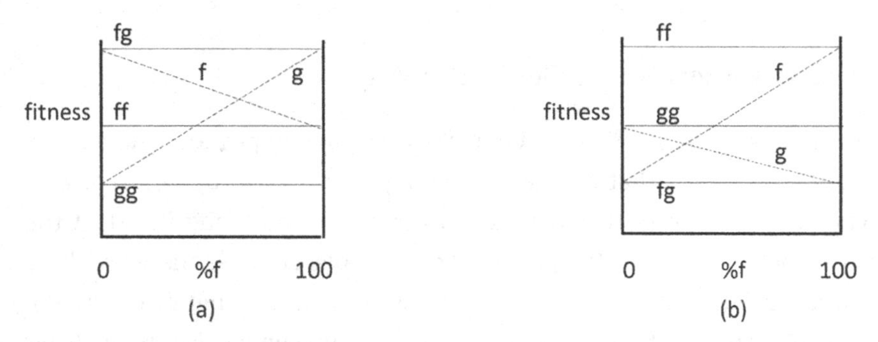

Figure 8.3 (a) Heterozygote superiority and (b) heterozygote inferiority.

that when selection is in control, f will evolve to 100 percent, and this will happen regardless of what the starting frequencies are of the two alleles; all that is needed is that both alleles are present in the population. The evolutionary outcome after numerous generations is that all the zebras in the population run fast.

Genotypes can be related to phenotypes in ways that alter the evolutionary outcome just described. Two such complications are depicted in Figure 8.3. In Figure 8.3a, heterozygotes are fitter than both the homozygous genotypes. To continue with the zebra examples, suppose that running speed is increased by lengthening the tibia, but the more the tibia is lengthened the more likely it is that the bone will break. In this instance, medium running speed may be optimal. Suppose that it is, and that medium running speed is caused by the heterozygote genotype. The broken lines in Figure 8.3a are the allelic fitnesses, and, as before, they are frequency dependent even though the genotypic fitnesses are not. Notice that f is fitter than g when f is rare and the reverse is true when g is rare. The result is that the population will evolve to an intermediate gene frequency, at which point all three genotypes (and all three phenotypes) are present in the population.[11] In this model, it is *impossible* for the fittest phenotype to evolve to fixation. Selection is an important cause of the evolutionary outcome, but it is not the only important cause.[12]

In Figure 8.3b, heterozygotes are less fit than both the homozygotes. For a simple example, consider the body size of a prey organism. If the organism is big, it can successfully fight back against predators; if it is very small, predators will leave it alone because small prey aren't worth chasing. The worst phenotype is to be middling in size. Suppose big and small are caused by homozygous genotypes and middling is caused by the heterozygote genotype. In this model, rare alleles are at a disadvantage.[13] Because of this, it is possible that the fittest genotype evolves to 100 percent, but it's also possible that the second-best genotype does so. Which outcome transpires depends on the population's starting frequency. This is also a case in which natural selection is not in control. Selection is an important cause of the evolutionary outcome, but it is not the only important cause.[14]

What I've said about the model in which there are two alleles at a locus can be extended to the case in which there are more than two, and similar

[11] When the population evolves to that equilibrium frequency, there still is selection on genotypes, but there is no change in gene or genotypic frequencies; this is an example of selection without evolution (Section 2.2). This provides a counterexample to Byerly and Michod's (1991) definition of the fitness of a genotype as its rate of increase in frequency (Maynard Smith 1991).

[12] If all you want to know is how the allelic frequencies will change, it suffices to attend just to their fitnesses; you can ignore the genotypic fitnesses and the fitnesses of the phenotypes that are associated with those genotypes. This resembles a point from Chapter 3; if all you want to know is how altruism and selfishness will change frequencies, it suffices to attend just to the average fitness of each. In both cases, it is a mistake to conclude that the factors you are ignoring are causally and explanatorily irrelevant.

[13] The allelic fitnesses when there is heterozygote inferiority resemble the stag hunt game described in Section 3.5.

[14] Terminology: When the heterozygote is exactly intermediate in fitness, there is no dominance; heterozygote superiority is called "over-dominance" and heterozygote inferiority is called "under-dominance." Note that genetic dominance is a different concept from that of drift dominating selection, which was discussed in Section 5.8.

	BB	Bb	bb
AA	3	2	1
Aa	2	0	0
aa	1	0	4

Figure 8.4 An example of epistasis. Cell entries are fertilities.

remarks also pertain to multi-locus models in which there is *epistasis*. Epistasis means that the fitness ordering of genotypes at one locus depends on the genotype present at another. An example is depicted in Figure 8.4. Assume that each two-locus configuration has its own phenotype. In this example, the fittest phenotype cannot evolve if selection is in control and the population begins with 100 percent A or with 100 percent B. Notice that there are two *adaptive peaks* in Figure 8.4, just as there are in the case of heterozygote inferiority depicted in Figure 8.3b.

Another way for genetic details to be a problem for ontological adaptationism concerns pleiotropy, a concept that is central to R. A. Fisher's geometric argument for gradualism (Section 6.5). The point of interest here concerns *antagonistic* pleiotropy. This occurs when a gene causes two or more phenotypes and at least one is beneficial and at least one is deleterious. G. C. Williams (1957) used this idea to construct a possible explanation of the evolution of senescence. Why do organisms often "fall apart" at the end of their lives? Do they do so because it helps the species to have evolution sweep out the old and make room for the new? As discussed in Chapter 3, Williams loathed hypotheses of group selection, so he felt compelled to explain senescence in some other way. He reasoned that a gene that benefits organisms early in life can evolve by natural selection even if it has deleterious effects later on. For example, an organism's fitness might not be diminished if the organism falls apart after its reproductive years are over and its offspring have been reared to independence. For Williams, senescence evolved because of natural selection, but pleiotropy was an important cause as well. Senescence is a byproduct of selection because there was no *selection for* senescence; the trait is not an *adaptation*. It resembles the colors of the balls in the toy depicted in Figure 2.1.

One of Gould and Lewontin's complaints about adaptationists is that they tend to "atomize traits." The objection is not that adaptationists identify single traits for biological investigation. That is inevitable and unobjectionable. Their point is that adaptationists often assume that traits evolve independently of each other. This independence assumption goes wrong when there is pleiotropy.[15]

8.4 Ancestral Influence (aka "Phylogenetic Inertia")

The Markov model for a dichotomous phenotypic trait discussed in Section 7.10 provides a handy device for thinking about what "ancestral influence" means in evolutionary biology. Recall that the Markov model entails that

(COR) Pr(descendant has trait T | ancestor has trait T) >
Pr(descendant has trait T | ancestor lacks trait T)

Ancestral influence is strongest when the time separating ancestor and descendant is short and attenuates as the temporal separation increases (Figure 7.9). The Markov model contains idealizations, however, so you have no *a priori* guarantee that what it entails is true. That said, suppose, for the sake of argument, that COR is true. That does not mean that ancestral influence is *powerful*, let alone that it is the *most powerful* cause of descendant phenotypes.

In view of the three-way distinction among propositions A, P, and O, consider the following remark of Roger Lewin's:

> Why do most land vertebrates have four legs? The seemingly obvious answer is that this arrangement is the optimal design. This response would ignore, however, the fact that the fish that were ancestral to terrestrial animals also have four limbs, or fins. Four limbs may be very suitable for locomotion on dry land, but the real reason that terrestrial animals have this arrangement is because their evolutionary predecessors possessed the same pattern (Lewin 1980, p. 886).

[15] The existence of sickle-cell disease in human populations where malaria is prevalent needs to be understood by seeing that the relevant genotypes affect both the disease and malaria resistance. The two phenotypic effects are not independent of each other.

Table 8.1 *Pr(most present-day land vertebrates are tetrapods | –)*

	MRCA was a tetrapod	MRCA was not a tetrapod
Stabilizing selection for tetrapody	w	x
No stabilizing selection for tetrapody	y	z

What does Lewin mean by "the real reason"? Is he asserting that ancestral influence was the only important cause, or, more modestly, that it was more important than selection, which was also important? Maybe ancestral influence and stabilizing selection both promoted the retention of tetrapody. What would it mean for one of these causes to be more powerful than the other? You've already seen an instance of this type of question – in the discussion in Section 5.8 of what it means for selection to dominate drift. The procedure used there – *the method of zeroing-out* – can be put to work here as well.

In Table 8.1, cell entries represent the probability that most present-day vertebrates are tetrapods conditional on different conjunctions. "MRCA" denotes the most recent common ancestor of present-day vertebrates. Horizontal inequalities ($w > x$ and $y > z$) represent the hypothesis that ancestral influence made a difference; vertical inequalities ($w > y$ and $x > z$) represent the hypothesis that selection made a difference (Orzack and Sober 2001). Let's assume that there was stabilizing selection for tetrapody, that the MRCA was a tetrapod, and that all these inequalities are true. The question before us is which of the causes was more powerful.

That question can be represented, I suggest, by asking whether $w - x > w - y$. Here the power of a cause is represented by how much difference the cause makes in the probability of the effect when that cause is varied while holding the other fixed. This inequality is true precisely when $y > x$. Evaluating whether $y > x$ requires you to consider what would have been the case if there were no stabilizing selection for tetrapody. Does that mean that limb number would have evolved by drift, or that there would have been selection for some limb number other than four? You also need to consider what would have been the case if the MRCA had not been a tetrapod. What number of limbs would it then have had? After these two questions are answered, the big question is whether $y > x$. If that question can't be answered, then Lewin's claim about "the real cause" is unjustified, but so is the claim that selection was "the real cause."[16]

The one-off character of Lewin's example may suggest that there's no way to tell whether selection was a more potent cause than ancestral influence, but other examples are possible in which the impact of selection and ancestral influence on evolutionary outcomes can be compared empirically. To illustrate that possibility, suppose that each of several bear species has either thick fur or thin, and that each species lives in an environment that is either cold or warm. Selection favors thick fur in cold climates and thin fur in warm. The selection hypothesis predicts that fur thickness should be correlated with climate. The ancestral influence hypothesis predicts that there should be a positive correlation between ancestral and descendant fur thickness. To see whether ancestral influence has a stronger impact on fur thickness than selection, you need to determine whether

Pr(species S has thick fur | S's ancestor has thick fur & S's environment is warm) >
Pr(species S has thick fur | S's ancestor has thin fur & S's environment is cold).

A data set of numerous lineages (as shown in Figure 8.5) might provide frequency data that allow you to test whether this inequality is true. Note that you need to know the character states of ancestors to do this.[17]

Lewin's idea that ancestral influence is more important than natural selection as a cause of tetrapody in land vertebrates is worth comparing with a more general remark of Darwin's that seems to entail the opposite judgment:

> It is generally acknowledged that all organic beings have been formed on two great laws: Unity of Type, and the Conditions of Existence. By unity of type is meant that fundamental agreement in structure which we see in organic beings of the same class, and which is quite independent of their

[16] **Exercise:** Analyze how this comparison of ancestral influence and stabilizing selection compares with the zeroing-out approach used in Section 5.8 to formulate a criterion for drift's dominating selection.

[17] Orzack and Sober (2001) discuss testing selection hypotheses and hypotheses of ancestral influence simultaneously, but do not consider the question of which cause is stronger. They use sister species rather than ancestors to construct the relevant "controlled comparison."

Figure 8.5 How often do present bear species have thick fur and how often do they have thin, in each of these two settings?

> habits of life. On my theory, unity of type is explained by unity of descent. The expression of conditions of existence, so often insisted on by the illustrious Cuvier, is fully embraced by the principle of natural selection. For natural selection acts by either now adapting the varying parts of each being to its organic and inorganic conditions of life; or by having adapted them during past periods of time: the adaptations being aided in many cases by the increased use or disuse of parts, being affected by the direct action of the external conditions of life, and subjected in all cases to the several laws of growth and variation. Hence, in fact, the law of the Conditions of Existence is the higher law; as it includes, through the inheritance of former variations and adaptations, that of Unity of Type. (Darwin 1859, p. 206)

Although Darwin's conclusion is that "conditions of existence" is more important than "unity of type," he notes that unity of type is a consequence of common ancestry and that the conditions of life are represented in his theory because they are part of the process of natural selection. That's why I think it's reasonable to interpret this passage as saying that selection is a more important cause than ancestral influence.

I expressed reservations about Lewin's claim about the primacy of ancestral influence, but that does not mean that I endorse Darwin's reasoning. To argue that ancestral influence is part of the explanation of why most present-day land vertebrates have four limbs, Lewin needs to say which ancestor he has in mind. Lewin does so – he is talking about the MRCA of present-day land vertebrates, not about ancestors that are more ancient. Lewin's claim would not be undermined if that MRCA had four limbs because of natural selection. Notice that Table 8.1 does not describe the possible causes of the MRCA's being a tetrapod. The fact that the evolution of ancient ancestors was strongly influenced by selection does not entail that lineages leading *from those ancestors to the present* followed suit.

In this section, I've used the term "ancestral influence" rather than another term that is sometimes used to label the same thing – *phylogenetic inertia*. I avoided the latter phrase because it misleadingly suggests that evolving lineages resemble billiard balls that continue to roll after they are pushed. If the antlers of the Irish elk initially increased in size because bigger antlers were fitter, did this cause the species to continue evolving in that direction after bigger antlers ceased to be advantageous, with the result that the species was driven to extinction? Some so-called "orthogenicists" in the early twentieth century embraced this line of thinking, but it was rejected in the Modern Synthesis shortly thereafter, and rightly so (Gould 1977b; Blomberg et al. 2003; Shanahan 2011).

8.5 Optimality Models

The genetic ideas reviewed earlier in this chapter may leave you wondering how methodological adaptationism could be a viable option. A partial answer is that there may be numerous examples of phenotypes in which the genetic scenarios I described do not apply. The fact of the matter is that evolutionary biologists often don't know what the underlying genetics are for the phenotypes they study. Should biologists abstain from considering hypotheses about the role played by natural selection until the full genetic story is made clear? The success of some optimality models suggests that this would be excessively timid.

Consider, for example, Geoffrey Parker's (1978) investigation of copulation time in yellow dung flies (Scatophaga stercoraria). When a fresh cowpat appears in a field, dung flies quickly colonize it, and males compete with each other for mates. After copulating with a female, a male spends time "guarding" her, meaning that he tries to prevent other males from mating with her. After that, he flies off in search of a new mate. What explains the male's guarding behavior? Given the opportunity, females will mate with multiple males. Parker found that the second male fertilizes far more eggs than the first. He discovered this by irradiating males with cobalt. Although irradiated sperm can fertilize eggs, the eggs do not develop. If an irradiated male copulates first and a normal male copulates second, about 80 percent of the eggs develop. If the mating order is reversed, only about 20 percent of the eggs develop. Parker concluded from this finding that a male gains a fitness advantage by preventing the female from mating further.

Parker wanted to explain the amount of time that yellow dung flies spend copulating. The observed average is thirty-six minutes. Parker found that longer copulation times are associated with more eggs fertilized, but there is a diminishing return on time invested – additional copulation time brings smaller and smaller increases in number of eggs fertilized. If copulation lasts for about 100 minutes, all the eggs are fertilized. The concave curve that Parker discovered is depicted in Figure 8.6. Another factor relevant to a male's reproductive success is that the time a male spends guarding one female is time not available for finding and copulating with another. This suggests that males do best when they copulate for less than 100 minutes. What copulation time would be optimal?

Parker found that the average amount of time that males spend searching for new mates and guarding them post-copulation is 156 minutes. This means that the three-part sequence of a male's reproductive behavior – search, then copulate, then guard – will last $156 + c$ minutes, where c is the amount of time spent copulating. To find the optimal value for c, you need to find the value of c that maximizes the number of eggs fertilized *per unit time.*

Parker represented the problem graphically (Figure 8.6). The x-axis represents the total time spent on the three tasks, and the y-axis represents the proportion of eggs fertilized. For any choice of c, you can calculate the number of eggs that will be fertilized. To find the maximum *rate* of eggs fertilized, you need to find the hypotenuse of a triangle. The hypotenuse of interest touches the fertilization curve and has the steepest slope. Once you find that hypotenuse, you can deduce the optimal value for c. The optimal value thus derived is forty-one minutes, which is close to the observed value of thirty-five minutes.

Parker's optimality model, narrowly construed, doesn't tell you how copulation time evolved. It just tells you what the optimal phenotype is. However, evolutionary biologists often are happy to add an historical interpretation to models like this. They view the optimal value as an *attractor* in parameter space – the species will evolve in the direction of that attractor, and once that parameter value is achieved, stabilizing selection will keep the species at that value. The assumption is that the fitness of a male's copulation time improves the closer it gets to the optimum.[18]

[18] Recall the similar assumption used by Fisher in his geometric argument (Section 6.5).

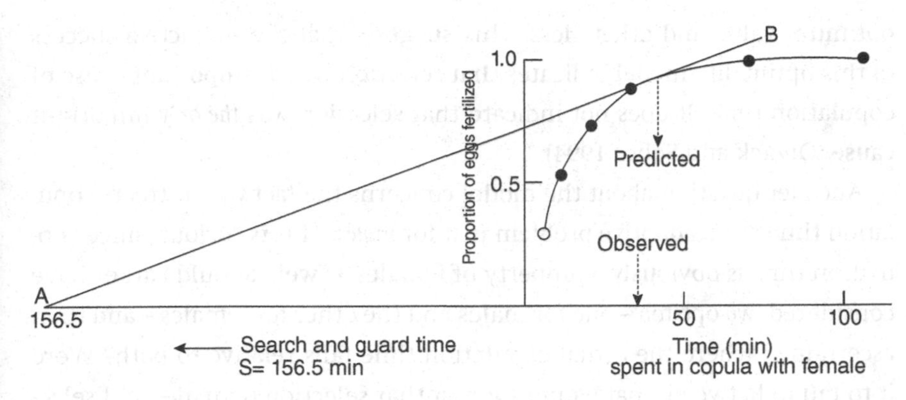

Figure 8.6 Yellow dung fly copulation time.

In their critique of adaptationism, Gould and Lewontin (1979) liken adaptationists to Dr. Pangloss, the cheerful character in Voltaire's *Candide* who repeatedly asserts that this is the best of all possible worlds even though his observations of the human condition are often hard to square with that rosy picture. Parker is no Pangloss. Parker inferred the optimal value for copulation time without assuming that the actual value is optimal or close to optimal. Parker's model leaves open the empirical question of how close the actual value is to the optimum.

Parker did not describe the genes in male dung flies that cause them to spend an average of thirty-five minutes copulating. Given that the model is genetics-free, it is striking how accurately it predicts the observed value. Models like this make *methodological* adaptationism attractive. However, what does this model tell you about whether *ontological* adaptationism is correct as a claim about copulation time in yellow dung flies?

One question concerns the six-minute gap between the optimal copulation time and the observed value. Should you say that this gap is negligible, and conclude that the model's predictive success shows that selection was the only important cause in the evolution of copulation time? Or does the gap indicate that some other cause was an important influence? Another issue arises from the fact that Parker was theorizing about *averages*. The curve in his model describes the *average* proportion of eggs fertilized as a function of copulation time, and 156 minutes is the *average* amount of time spent searching and guarding. Individuals differ from each other with respect to both these phenotypes. If males on average are six minutes away from the forty-one-minute optimum, some males deviate more from the optimum value and others less. This suggests that the predictive success of this optimality model indicates that selection was *an* important cause of copulation time. It does not indicate that selection was *the only* important cause (Orzack and Sober 1994).

Another question about the model concerns the fact that it treats copulation time as an adaptive problem just for *males*. This is curious, since copulation time is obviously a property of females as well. Should Parker have considered *two* optima – one for males and the other for females – and then ascertained where the actual copulation time falls relative to both? Were it to fall in between, that would suggest that selection on males and selection on females both influenced the outcome. Indeed, the fact that females benefit from multiple matings and seek new mates when left unguarded suggests that the female optimum is less than the male optimum. If the observed copulation time is a compromise between what would be optimal for males and what would be optimal for females, it isn't true that the observed copulation time is optimal full stop.[19] This reasoning may help explain the six-minute gap.

Although Parker sees copulation time as a male adaptation, there are other aspects of his study that suggest how females have evolved adaptations for dealing with males.[20] Arguably, the concave curve that plots proportion of eggs fertilized as a function of copulation time is a product of selection on females. If a female ensures that it takes 100 minutes of copulation for a male to fertilize all her eggs, this puts selection pressure on males to stop copulating well before all her eggs are fertilized. The result is that females are able to gather sperm from multiple males; the resulting sperm competition allows her sons to be fitter, since her sons will do better in their reproductive careers.

[19] The example of zebra tibia in Section 8.3 shows that some compromises aren't a problem for adaptationism. Dung fly copulation time might be a different story. Here's another example: if group selection favors altruism and individual selection favors selfishness, you can see whether the trait frequencies in the metapopulation are closer to 100 percent altruism or to 100 percent selfishness. If the observed frequency in the metapopulation is significantly different from both, the outcome isn't optimal for the group or for the individual.

[20] In a later paper, Parker et al. (1999, p. 804) say that "our results show that although males adjust copula duration to maximize their own fitness, they do so because of selection imposed by female morphology."

My last point about Parker's model concerns the disdain that anti-adaptationists frequently express for optimality models. These critics often claim that optimality models assume that organisms are optimal, and they deride the models' idealizations. Some of this negativity makes sense if the point is simply that additional models should be constructed and tested. However, this does not mean that optimality models are useless. The reason is simple: if you want to reject the claim that natural selection was the only important cause of the evolution of a trait, you need to know what trait would evolve if natural selection were the only important cause. *Adaptationists and anti-adaptationists both need optimality models* (Orzack and Sober 1994).[21]

8.6 The Chronological-Order Test of Adaptation Hypotheses

I just described how optimality models can be used to test adaptationist claims about specific traits in specific populations. A very different strategy is available for testing the more modest claim that a given type of natural selection was an important cause of a trait's evolution. It brings in phylogenetic considerations. Darwin pioneered this approach, though he did not spell out the logic of his argument in full detail.

Here's an example.[22] Mammals in utero have skull sutures that facilitate live birth. The question is whether the sutures evolved *because* they facilitate live birth. Here is Darwin's answer:

> The sutures in the skulls of young mammals have been advanced as a beautiful adaptation for aiding parturition, and no doubt they facilitate, or may be indispensable for this act; but as sutures occur in the skulls of young birds and reptiles, which have only to escape from a broken egg, we may infer that this structure has arisen from the laws of growth, and has been taken advantage of in the parturition of the higher animals (Darwin 1859, p. 197).[23]

[21] Recall Galton's point about the need for a baseline in explaining the extinction of family names (Section 4.1).

[22] For other examples, see Sober (2011b, chapters 1 and 3).

[23] Darwin found this example in Richard Owen's (1849, p. 145) book, *On the Nature of Limbs*. Owen used it to argue that some traits of organisms need to be understood in terms of their relation to an *archetype*, not in terms of their current functional utility. Darwin replaced Owen's "archetypes" with *ancestors*.

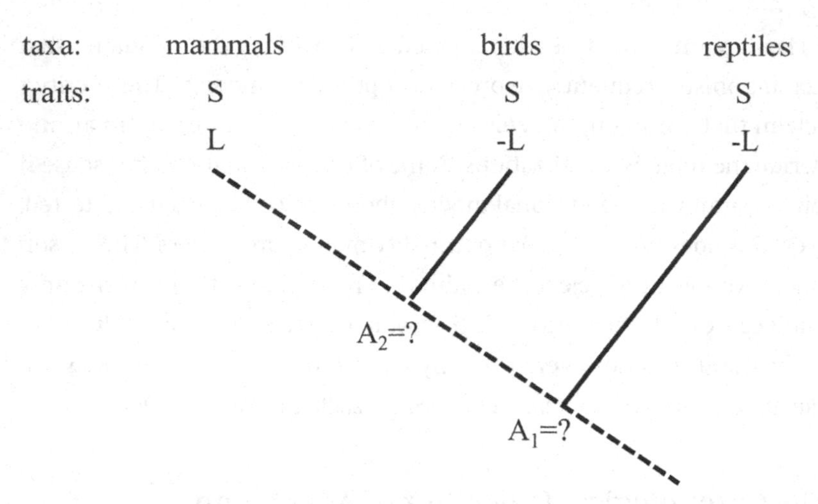

Figure 8.7 Darwin's phylogenetic reasoning about the evolution of skull sutures (S) and live birth (L) in mammals.

This passage may strike you as puzzling. Why did Darwin think that birds and reptiles are relevant, given that his question concerns mammals?

The answer, I suggest, is that Darwin was thinking about a phylogenetic tree like the one depicted in Figure 8.7.[24] This tree and the observed traits of its leaves allow you to infer the character states of ancestors A_1 and A_2. Darwin inferred that A_1 and A_2 both had skull sutures, but neither had live birth. It follows from those conclusions that the lineage leading to current mammals (depicted by the broken line) had skull sutures long before live birth evolved. This means that it is false that skull sutures are *adaptations* for parturition.

Darwin did not justify his assignment of character states to ancestors A_1 and A_2, but he had the resources to do so. His strong principle of inheritance (Section 1.6) – that offspring tend to resemble their parents – could have done the job if his tendency claim is interpreted in terms of the COR inequality (Section 7.10). Cladistic parsimony puts its stamp of approval on this inference as well (Coddington 1988).

The chronological-order test is relevant to the discussion in the previous section of Parker's (1978) model of dung fly copulation time. Parker

[24] Cladists will smile (or wince) at Darwin's putting reptiles on the leaves of this tree. **Questions:** Why should they wince? How can Figure 8.7 be modified to meet their objection, and how would that modification affect Darwin's conclusion?

uses the observed amount of time that males spend searching+guarding to infer what the optimal copulation time is. When the model is given an evolutionary interpretation, this means that the model is saying that searching+guarding time evolved first, with copulation time evolving subsequently as an adaptive response. This assumption about chronological order needs to be justified. Phylogenetic information may be helpful here.[25]

8.7 Testability and Research Programs

Specific optimality models can be tested, and so can specific selection claims about why a trait evolved in a lineage. That much is clear from Parker's model of dung fly copulation time and Darwin's discussion of mammalian skull sutures. But what do such tests tell you about the big-picture claim made by ontological adaptationism? Is it testable too?

Gould and Lewontin (1979. p. 586) seem to suggest that it is not when they say that

> … the adaptationist programme can be traced through common styles of argument … : (1) If one adaptive argument fails, try another … (2) If one adaptive argument fails, assume that another must exist … (3) In the absence of a good adaptive argument in the first place, attribute failure to imperfect understanding of where an organism lives and what it does.

My response is that it is important to recognize the distinction between a *proposition* (ontological adaptationism) and the *people* who endorse that proposition. Even if adaptationists are sometimes, usually, or always dogmatic, that says nothing about whether the proposition is testable. Young-earth creationists may be dogmatic in their claim that life on earth began no more than 50,000 years ago, but that doesn't show that the claim is untestable. It *is* testable; otherwise, scientists would be in no position to say that it is false.

[25] **Question:** I mentioned before that Gould and Lewontin (1979) chide adaptationists for being soulmates of Dr. Pangloss, the arch naïve adaptationist. A humorous feature of Voltaire's satire is that Pangloss consistently gets things backwards, as when he says that "the nose is formed for spectacles, therefore we wear spectacles. The legs are visibly designed for stockings, accordingly we wear stockings." Can Darwin's analysis of mammalian skull sutures be used to show that Pangloss's pronouncements are mistaken?

That said, there is an important difference between specific adaptationist hypotheses and ontological adaptationism, which is about all the phenotypes of all living things on earth. The latter claim is testable, but only in the long-run (Orzack and Sober 1994). It is only by testing a host of specific hypotheses that facts about the big picture can fall into place. Consider the analogous problem of whether speciation is almost always allopatric, meaning that it proceeds by first having a population of conspecifics broken apart by a geographical barrier, after which the separated daughter populations evolve traits that render them reproductively isolated from each other. Sympatry is an alternative hypothesis; it says that speciation occurs without an initial geographical separation. The way to discover which of these two speciation processes was more prevalent is to study speciation events one-by-one. Discussions of adaptationism, both *pro* and *con*, often give the impression that biologists need to commit to one of these positions *before* they study specific traits in specific populations. This imperative has things backwards.

Although ontological adaptationism is a proposition about nature, methodological adaptationism is a research program, so it is useful to put that idea in a larger context. Karl Popper (1959) wanted to solve the demarcation problem of separating science from non-science. His solution was to talk about *propositions*; for Popper, a proposition is scientific if and only if it is falsifiable. Problems with Popper's formulation[26] led Imre Lakatos (1978) to think about the demarcation problem in a different way. He wanted to separate progressive from degenerative research programs. Lakatos suggested that you can judge whether research programs are progressive or degenerative by seeing how they manage to solve problems. Progressive programs make novel predictions that turn out to be true; degenerative

[26] I'll mention two problems. Here is Popper's definition of falsifiability: a proposition is falsifiable precisely when it deductively entails at least one observation statement. The first problem with this idea is that probability statements are unfalsifiable. The hypothesis that a coin is fair does not deductively entail how many heads you'll get in a sequence of tosses; the hypothesis is compatible with all possible observational outcomes. However, it's absurd to say that this and other probability statements are unscientific. The hypothesis about the coin is *testable*, but it isn't *falsifiable* in Popper's strict sense of the term. Another problem is that Popper's criterion entails that definitional truths are unscientific. This also is absurd. Surely the definition of effective population size in population genetics is a scientific proposition.

programs manage to accommodate old evidence ad hoc, but never issue in successful novel predictions.[27]

Lakatos's idea is fruitfully supplemented by Philip Kitcher's (1990) discussion of "the division of cognitive labor." Kitcher thinks that science thrives when different research groups, all working on the same problem, use different research strategies. Even if there is a single best approach to solving a given problem, it would be a mistake to have all research groups adopt it. Better to have some take the best approach, others take the next best approach, and so on, with resources allocated to approaches according to how promising they are. The result is that scientists who take the best approach have lots of competitors who are doing the same, while those who take the second-best approach have fewer. Individual researchers don't just strive to solve a problem; they strive to solve it *first*. This selfish goal of individuals leads to diversity among scientists, and that diversity serves the goals of science.[28] Kitcher's idea about the optimal division of cognitive labor among research groups is relevant to judging the status of adaptationism as a research program. The question is not whether *all* evolutionary biologists should embrace methodological adaptationism or whether *none* should do so.

[27] Mitchell and Valone (1990) apply Lakatos's idea to what they call "the optimization research program" in evolutionary biology.

[28] Kitcher's idea echoes Adam Smith's invisible hand, as does Hull's (1988) account of the scientific process.

9 Big-Picture Questions

9.1 Are There Laws of Evolution?

The final sentence of the first edition of the *Origin* raises a philosophical question:

> There is grandeur in this view of life, with its several powers, having been originally breathed into a few forms or into one; and that, whilst this planet has gone circling on according to the fixed law of gravity, from so simple a beginning endless forms most beautiful and most wonderful have been, and are being evolved (Darwin 1859, p. 490).

Darwin's mention of the breath of life is noteworthy,[1] but what interests me here is this: if there are fixed laws for planetary revolution, are there also fixed laws for biological evolution? Darwin wanted to be the Newton of biology; he wanted to find laws that explain the similarities and differences we observe organisms to have, just as Newton found laws that explain celestial and terrestrial motion. Did Darwin actually do so?

In Newtonian physics, gravity explains why the earth revolves around the sun, and Newton characterizes that force by his universal law of gravitation:

$$F_g = G\frac{m_1 m_2}{r^2}$$

This equation says that the force due to the gravitational attraction that two objects exert on each other is equal to the gravitational constant times

[1] The breathing is Biblical: "Then the Lord God formed man of the dust of the ground, and breathed into his nostrils the breath of life; and man became a living soul (*Genesis* 2:7)." Just to make this point clear, Darwin added "by the Creator" in the second edition of the *Origin*. See Sober (2011b, Chapter 4) for discussion of this and other passages in connection with the question of whether Darwin embraced methodological naturalism.

the product of their masses divided by the square of the distance between them. Newton's law of gravitation has the several characteristics that scientists and philosophers often use to define what a law of nature is. The standard picture I have in mind says that laws are true empirical generalizations that don't mention any token individual, place, or time, and that support counterfactuals.[2] This last requirement is satisfied by Newton's law, since that law covers all *possible* values for mass and distance, most of which fail to be exemplified by objects that actually exist. The law thereby makes pronouncements about possible states of affairs; laws aren't limited to describing what actually happens, though they do that too.[3]

The parallel to gravitation in Newton's theory is natural selection in Darwin's, but what is the law that characterizes the nature of that force? For the sake of a simple answer, let's consider the following, though Darwin, as far as I know, never put things exactly this way:

> The intensity of selection on trait T in a population at a given time = the amount of variation in fitness there is among the different trait values of T that are exemplified in the population at that time.

Here T is a variable (like height) that can take different values. One difference between the law of gravity and this statement about selection is that Newton's law tells you the unique physical basis (or source) of gravitational attraction – the masses of objects and the distances between them. Fitness, on the other hand, is *multiply realizable* (Section 2.2). This difference, by itself, does not show that the displayed proposition about selection is not a law, but there is another feature of this example that most philosophers think does just that. The statement about selection is not empirical; it is true by definition. This is where I dissent from the traditional conception of laws. I think that laws can be a priori, standard examples like Newton's law of gravitation notwithstanding.

The examples I have in mind are dynamical models in evolutionary biology – models that describe how a population will change through time.

[2] A true generalization of the form "All As are Bs" supports a counterfactual precisely when "for any x, if x were an A, then x would be B" is also true. Generalizations that don't support counterfactuals are said to be *accidental generalizations*. An example is "all gold objects on earth weigh less than ten tons."

[3] For a very different conception of laws, see the Humean "best systems account" of laws proposed by David Lewis (1973, 1994).

Table 9.1 *A one-locus two-allele model of viability selection*

	Genotypes		
	AA	Aa	aa
Frequency before selection	p^2	$2pq$	q^2
Fitnesses	w_{AA}	w_{Aa}	w_{aa}
Frequency after selection	$\frac{p^2(w_{AA})}{\bar{w}}$	$\frac{2pq(w_{Aa})}{\bar{w}}$	$\frac{q^2(w_{aa})}{\bar{w}}$

Consider an example from the previous chapter (Figures 8.3) – the one-locus two-allele model of selection where genotypic fitnesses are frequency independent. Table 9.1 represents this model when there is selection only on viability. The fitnesses are probabilities of surviving to adulthood. The three frequencies before selection sum to 1 (since $p + q = 1$), and so do the frequencies after selection, since $\bar{w} = p^2(w_{AA}) + 2pq(w_{Aa}) + q^2(w_{aa})$; $\bar{w}$ is the average fitness of the organisms in the population (Section 2.8). This table provides a one-generation snapshot of how differences in viability lead to a shift in genotypic frequencies as embryos develop into adults. If there is random mating after selection, you can figure out from the parental frequencies what the genotypic frequencies are in the embryos of the next generation. You then can put those frequencies into a new table and apply the same calculation to see how frequencies will change in that generation, again owing to differences in viability. By looping through multiple applications of this algebra, you have a simple model of how selection works in a sequence of generations.

This model has an interesting consequence when there is heterozygote superiority. The alleles in the population will evolve to an intermediate frequency that depends just on the three genotypic fitnesses. The intermediate frequency of the *A* allele towards which the population will evolve is:

$$\hat{p} = \frac{w_{Aa} - w_{aa}}{(w_{Aa} - w_{aa}) + (w_{Aa} - w_{AA})}.$$

You can derive this equation from the allelic fitnesses

$$w_A = p(w_{AA}) + q(w_{Aa}) \qquad w_a = q(w_{aa}) + p(w_{Aa})$$

by using the fact that these fitnesses are equal when the population achieves its equilibrium frequency.

Table 9.2 *Two explanations – one using an empirical law, the other not*

	Newtonian Physics	Evolutionary Biology
Initial conditions	Masses of earth and sun and distance between them. Other forces have negligible effects.	Fitnesses of AA, Aa, aa in a population where Aa is fittest and there is random mating. Other forces have negligible effects.
Laws	universal law of gravitation and F = ma	1-locus 2-allele model of heterozygote superiority
What is explained	Earth and sun accelerate towards each other.	Allelic frequencies evolve towards $\hat{p}$.

Now consider a conditional. Its antecedent describes all the assumptions in this model, along with the assumption that selection as described in the model is in control of the evolution of the three genotypes. The conditional's consequent says that the population will evolve towards the equilibrium frequency $\hat{p}$. This conditional is a mathematical truth. Like the claim that 2 + 3 = 5, it is a priori. Don't be misled by the fact that the conditional's antecedent is empirical and so is its consequent. Similarly, "If Joe is bachelor, then Joe is unmarried" is a priori, but its antecedent and consequent are not.

The conditional just described is a simple model drawn from evolutionary biology 101, but more sophisticated models have the same feature. For example, the Kimura model that describes how the fixation probability of a token mutation is determined by its selection coefficient and the effective population size (Section 5.6) is also a priori.[4]

If dynamical models in population biology are often a priori true, why say that they express laws of nature? Why fly in the face of established tradition, which insists that laws of nature must be empirical and can't be mathematically necessary? The reason I am a rebel is that these mathematical models are exactly like empirical laws of nature in the way they figure

[4] I endorse a claim about causation that parallels the above claim about laws. I think there are claims of the form "*a*'s being *F* would cause *b* to be *G*" that are a priori true and genuinely explanatory (Sober 2011a). Lange and Rosenberg (2011) argue to the contrary, to which Elgin and Sober (2015) reply. For discussion of whether "F = ma" is a priori, see Earman and Friedman (1973).

in explanations and predictions (Elgin 2003; Sober 2011a). To see why, consider the two explanations described in Table 9.2. One uses Newton's law of gravitation (which is a source law) and F = ma (which is a consequence law); the other uses the 1-locus 2-allele model of heterozygote superiority. In each explanation, you explain why a singular proposition is true by deducing it from a statement describing initial conditions and a statement describing a law. The key property of dynamical laws that allows them to contribute to explanations is that they link earlier causes (the initial and boundary conditions) to their later effects. The link can be forged by an empirical proposition or by an a priori truth.[5]

Dynamical laws of evolution that are mathematical truths are counterexamples to the reductionist thesis that every law of nature that isn't a law of physics can be explained by physics. Physics explains these dynamical laws no more than it explains why 2 + 3 = 5. Even if organisms are made of matter and nothing else, it does not follow that the laws of biology reduce to the laws of physics. The result is that biology has a kind of *autonomy*.[6,7]

[5] I am not claiming that evolutionary biology is the only science that contains a priori dynamical laws. Diez and Lorenzano (2015) argue that it is easy to construct a priori laws from the empirical laws found in other sciences, and that there is therefore no difference between the situation in evolutionary theory and that in physics. I agree with their first point, but disagree with the second.

[6] Fodor's (1974) very different argument for the autonomy of the "special sciences" has been influential; for criticisms, see Sober (1999).

[7] **Exercise:** My argument that dynamical laws in evolutionary biology are often a priori leaves open whether others are empirical, and it says nothing about the status of non-dynamical laws. Are there any empirical biological laws? John Beatty (1995) argues that no empirical biological generalizations are laws. To develop his argument, he uses the traditional idea that laws can't be accidental generalizations. His thesis is that true empirical biological generalizations are always accidents of history. This means they lack the kind of necessity that laws are required to have. For example, consider the Mendelian law of segregation, which says that AB heterozygotes produce A and B gametes with equal frequency. This "law" isn't true when there is meiotic drive (Section 3.10). Now consider a population in which there is no meiotic drive and gamete frequencies conform to the Mendelian principle. Beatty's point is that this generalization about the population is true because of an historical accident – that there are no driving genes in the population. So even if the Mendelian generalization were true, it would not be a law. I think Beatty is right about this example, but the question remains about whether there are counterexamples to his general thesis. To explore this question, check out "Biological Rules" in Wikipedia for some examples to consider; also see Sober (1997) for a reply to Beatty's argument.

9.2 Determinism, Indeterminism, and Objective Probabilities

Ernst Mayr (2007, p. 111) says that the theory of natural selection dethrones determinism. He doesn't explain what he means, but I suspect he was struck by the fact that selection involves probabilities whereas determinism involves necessities. In population genetics it is more standard to say that selection is deterministic, meaning that when it is in control of the evolutionary process, there is a single outcome that must happen (which our rule of thumb describes). In this usage, indeterminism arises from finite population size, which means that there are multiple possible futures, each with its own probability, given the population's initial condition.

The definition of "determinism" usually used by physicists and philosophers is different. For them, determinism asserts that a *complete* specification of the initial conditions and the boundary conditions of a system at a given time ensures that there is just one possible future; it's the one that must come to pass (Earman 1986).[8] The system's *initial* conditions are the ones that obtain at a time of the investigator's choosing; its *boundary* conditions are the conditions outside of the system that may impinge on the system later on. Indeterminism is the thesis that a *complete* specification ensures that there are at least two possible futures. It follows from these definitions that you can't conclude that coin tossing is an indeterministic process from the fact that you have lots of evidence that

$$0 < \Pr(\text{the coin lands heads at time } t_2 \mid \text{the coin was tossed at time } t_1) < 1.$$

The statement "the coin was tossed at t_1" is true but it is an incomplete description of what was true at that time. The same point applies to the fact that *models* in evolutionary biology use probabilities; this, by itself, is no proof that evolutionary *processes* are indeterministic.

The concept of completeness used here can be clarified by using the concept of screening-off (Section 4.4):

> If C describes the state of the system at t_1 and E describes its state at t_2, then C at t_1 is complete relative to E at t_2 precisely when Pr(E at t_2 | C at t_1) = Pr(E at t_2 | C at t_1 & H) for any H that describes the history of the system up to and including t_1.

[8] The definition just given describes *forward-directed* determinism. A similar definition can be defined for backward-directed determinism.

This equality says that C at t_1 screens-off H from E at t_2. Note that the completeness of C at t_1 relative to E at t_2 does not mean that C includes all the propositions that are true at t_1. The probabilities used here are *objective*, by which I mean that it isn't facts about the degrees of belief that agents have or should have about the relationship of the conditioned and the conditioning propositions that make the equality true or false.

With this clarification in hand, I want to consider whether determinism entails that there are no objective probabilities that have nonextreme values. The thought that determinism has this consequence is suggested by a famous passage from Pierre-Simon Laplace's (1812) *Philosophical Essay on Probabilities*:

> We may regard the present state of the universe as the effect of its past and the cause of its future. An intellect which at a certain moment would know all forces that set nature in motion, and all positions of all items of which nature is composed, if this intellect were also vast enough to submit these data to analysis, it would embrace in a single formula the movements of the greatest bodies of the universe and those of the tiniest atom; for such an intellect nothing would be uncertain and the future just like the past would be present before its eyes.

Inspired by Newton's physics, Laplace is assuming that determinism is true. This is why the hypothetical "intellect" that Laplace describes (now known as Laplace's "demon") does not need to use probabilities to predict what will happen. Why, then, do we humans find probabilities indispensable? The answer is that we have incomplete knowledge of the current state of the universe and we are limited in our abilities to calculate. According to Laplace, it is facts about our minds that explain why we use probabilities. Does it follow that there are no objective probabilities if determinism is true? The answer is *no*, and the distinction between pragmatics and semantics explains why. Pragmatics is the study of how people use language; semantics is the study of what the sentences in a language mean. I might use the English sentence "you are kind" to express my gratitude, but the sentence does not mean that I am grateful. The fact that we *use* a probability statement because we are ignorant does not entail that the statement *describes* our ignorance.

This reply to the Laplacean argument does not indicate where objective probabilities come from if determinism is true. To explore that question,

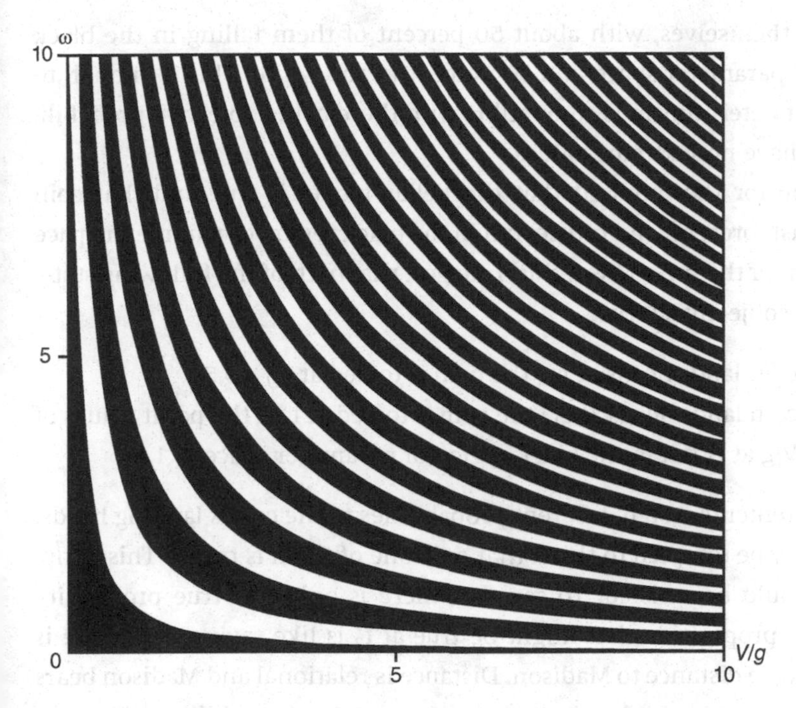

Figure 9.1 A Newtonian model of coin tossing.

let's consider a simple Newtonian model due to Joseph Keller (1986) that describes coin tossing as a deterministic process. There are two causal factors that jointly determine whether the coin lands heads or tails. There is the upward momentum (w), which determines how high the coin will rise, and the angular momentum (V/g), which describes how fast the coin will spin. The model contains some simplifying assumptions: the coin is always tossed from the same height, it lands on sand so there is no bouncing, there are no breezes, the coin spins along its axis, and the coin starts heads-up on your thumb. A specification of values for the upward and angular momenta locates your toss at a point in the square depicted in Figure 9.1, which comes from Diaconis (1998). If you give the coin very little angular momentum, it will rise like a flying saucer and then land heads. If you give it very little upward momentum, it will hardly rise at all, so heads will be the result. Either way, the coin will land heads because its pair of values is located in the L-shaped black region of the figure that spreads out from the origin. Given this deterministic model, repeated tossing of a coin produce around 50 percent heads because your tosses differ amongst themselves, with about 50 percent of them falling in the black region of parameter space and 50 percent in the white. This is enough to show that determinism is compatible with there being objective probabilities that have nonextreme values.[9]

Assume for the moment that the Keller model is true and that coin tosses (past, present, and future) fall in the black region of parameter space 50 percent of the time. Given this, I suggest that both of the following statements are objectively true:

- Pr(the coin lands heads at t_2 | the coin is tossed at t_1) = 0.5
- Pr(the coin lands heads at t_2 | the coin is tossed at t_1 & the point values of w and V/g at t_1 fall in the black region of parameter space) = 1.0

These statements assign different probabilities to the coin's landing heads, so you may be tempted to think that only one of them is true.[10] This temptation should be resisted. To say that there is only one true probability at t_1 for a proposition that might be true at t_2 is like saying that there is only one true distance to Madison. Distance is relational and Madison bears that relation to multiple *relata*; the same goes for probability.[11] The two bulleted probability statements conditionalize on different propositions; they are logically compatible. The *principle of total evidence* says that if you want to predict at t_1 what will happen at t_2, you should use all the relevant information you have. In the case at hand, the principle recommends using the second probability to make your prediction if you know that the point values of w and V/g fall in the black region of parameter space. However, this principle is part of the pragmatics of probability. That has nothing to

[9] Given what Lewis (1980) means by "chance," he is right that objective chances must have values of 1 or 0 if the universe is deterministic. For Lewis, the chance at t_1 that a proposition H will be true at t_2 is the probability that H has, conditional on the whole history of the world up to and including t_1. Gillies (2016) attributes a similar view to Popper (1990) and Miller (1994), who hold that the probabilistic propensity a hypothesis has of being true at a given time needs to be based on a complete description of the state of the universe before that time. To me, these are idiosyncratic uses of the terms "chance" and "probabilistic propensity," but the important point is that these theses do not conflict with my claim that nonextreme probabilities can be objective in a deterministic world. See Schaffer (2007) for an argument to the contrary.

[10] Hempel (1965) would call this an example of "statistical inconsistency."

[11] There are two exceptions to the thesis that probability is relational. Tautologies have probabilities of 1 and contradictions have probabilities of 0, period.

do with the nonpragmatic question concerning which of those probability statements is true. Both are. This point about the relational character of probability also holds if determinism is false.

9.3 Must Micro-indeterminism Percolate Up?

If nature is indeterministic at the level of elementary particles,[12] does that entail that evolutionary processes at the level of genes, organisms, and populations are also indeterministic? I'll explore this question by using the concepts of *supervenience* and *multiple realizability* (Section 2.2). Figure 9.2 resembles a famous figure that Fodor (1974) used to argue against reductionism (Section 3.13); it depicts two macro-event types (P and Q) that are each multiply realizable at the micro-level. The A's are mutually exclusive as are the B's; they capture all (and only) the micro-supervenience bases that P and Q have. I'm assuming that the As and the Bs are finite in number. Each A_i, if it occurs at t_1, ensures that P occurs at t_1, and each B_j, if it occurs at t_2, ensures that Q occurs at t_2. For a macro-event to occur, one of its micro-supervenience bases must occur. Suppose that each A_i confers probabilities on each of the B_js, and these probabilities are all less than 1. Let's further suppose that the As are complete with respect to the Bs, so the relationship of the As to the Bs is genuinely indeterministic. What does this entail about whether $Pr(Q \text{ at } t_2 \mid P \text{ at } t_1) < 1$?

When there is multiple realizability and supervenience, micro-*indeterminism* entails macro-*determinism* precisely when the following equality holds:

$$\text{For each } A_i, \Sigma_j \Pr\left(B_j \text{ at } t_2 \mid A_i \text{ at } t_1\right) = 1$$

If P happens at t_1, then one of the A_i's must happen at t_1, and the above equality then guarantees that one of the B_j's must happen at t_2, in which case Q must happen at t_2. This shows that micro-indeterminism does not entail macro-indeterminism. However, I think the above equality rarely if ever holds; it is almost always true that at least one of the

[12] This is suggested (but not proven) by quantum mechanics, as there are indeterministic interpretations of quantum mechanics as well as deterministic interpretations, and they all fit the available observations. These so-called "interpretations" are distinct theories, since they have distinct ontologies (Maudlin 2019). My thanks to Eddy Chen for clarification on this point.

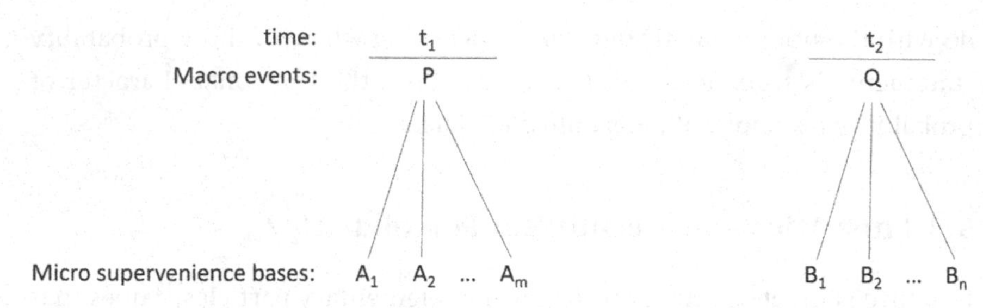

Figure 9.2 Does micro-indeterminism entail macro-indeterminism?

micro-supervenience bases of P at t_1 confers a non-zero probability on at least one of the micro-supervenience bases of *not*Q at t_2. In that circumstance, micro-indeterminism entails macro-indeterminism (see Glymour 2001).[13] This conclusion does not rule out "near-determinism" at the macro-level, meaning that Pr(Q at t_2 | P at t_1) is very close to 1.0 (Glennan 1997; Graves et al. 1999). However, near-determinism is not a kind of determinism, any more than nearly winning a race is a way to actually win it.

My argument does not concern whether events at the micro-level can *cause* events at the macro-level, or if they can, how often they actually do so (on which see Millstein 2002b). However, I suggest that *they can and always do*. I think that if a P event at time t_1 causes a Q event at t_2, then the instantiation of one of the As at t_1 also causes that outcome. For example, if Jill's inhaling cigarette smoke caused her to get lung cancer, then the interaction of her cells with the molecular constituents of smoke caused her to get lung cancer (Shapiro and Sober 2007).

9.4 Objective Interpretations of Probability

If probabilities are objective quantities in evolutionary theory, how should those probabilities be interpreted? They don't describe the degrees of belief that an agent has or ought to have, so what do they describe?[14] There

[13] I thank Andrew Cuda for spotting a flaw in an earlier formulation of this argument.

[14] If "Pr(Y | X) = p" expresses an objective relationship between X and Y, agents who know this probabilistic proposition may want to adjust their degrees of belief so that Cr(Y | X) =p (where "Cr" denotes subjective credence). The connection just drawn between objective probabilities and credences resembles Lewis' (1980) *principal principle*, but note that here I am talking about objective probabilities, not about Lewisian chances. My formulation may be a bit more useful, since agents rarely know the

is a longstanding philosophical literature on objective interpretations of probability (ably analyzed by Galavotti 2015 and Hájek 2019). The usual suspects are actual frequencies, hypothetical frequencies, and propensities. An acceptable interpretation must satisfy the mathematical axioms of probability, but it is a further question whether and where each gets used in evolutionary biology.

Before turning to evolutionary biology, I want to take two formulations of the hypothetical-frequency interpretation off the list of possibilities. Consider these two claims about a fair coin:

- If you toss the coin again and again, the frequency of heads will get closer and closer to 50 percent.
- For any $e > 0$, Pr(frequency of heads falls in the interval $[0.5 + e, 0.5 - e]$ | n tosses) $\to 1$ as $n \to \infty$.

The first statement is false. Fair coins are not obliged to march monotonically towards 50 percent heads; in fact, they can't. The second statement is true; it expresses a version of the *law of large numbers*. However, it isn't an *interpretation* of what it means for a coin to have a probability of landing heads of 0.5, since the law uses the very concept one wants to elucidate. An interpretation of probability must use a concept that is already clear, in terms of which probability can be understood.

The actual frequency of a trait in a population is an important parameter in evolutionary models. Evolution in such models means change in trait frequencies.[15] These frequencies are probabilities since they satisfy the axioms of probability. The unconditional probability that an individual in a population at a given time has trait T is simply the frequency of that trait in that population at that time.

In contrast, the viability fitness of trait T is a conditional probability – Pr(o survives to adulthood | o has trait T) – that needn't be identical with the frequency with which individuals that have trait T survive to adulthood (Section 2.2), so how should that probability be interpreted? It is here that

Lewisian chances of future events. Note also that I do not say that rational agents are *obliged* to have their subjective credences match these known objective probabilities; they may have other information that is relevant – for example, they may already know that Y is true. This is a possibility that Lewis recognizes in connection with his principal principle.

[15] Though recall the *caveat* from Section 1.3.

philosophers have reached for the propensity interpretation of probability. I take the propensity interpretation of probability to be saying that "Pr(Y | X)" describes the degree to which X's being true causally promotes Y's being true. The problem with this proposal is that the causal relationship between X and Y is not well-represented by a single probability. The probability of your living to age 100 if you read philosophy books is small, but that isn't because reading philosophy is bad for your health. At the minimum, a claim about causal strength must describe how much *difference* reading philosophy books makes in longevity; this involves *comparing* probabilities.[16]

This point shows that the propensity interpretation of probability has a problem quite apart from the question of whether it applies to the concept of fitness. However, there are facts about the fitness concept itself that also suggest that the fitness of a trait does not represent the causal power of the trait to promote survival and/or reproductive success. Consider the selection toy (Figure 2.1); coextensive traits can differ in their causal powers, though they have the same fitness value.

Although the fitness of a trait does not represent the trait's causal influence on survival and reproduction, fitness *differences* are another story. When a mutation is introduced into a population that has the wild type gene W at fixation, the selection coefficient s is the mutation's fitness minus the fitness of W (Section 5.2). This difference is a good representation of a causal property that the mutation has in that context; it represents a causal factor that affects how the population will evolve. However, the selection coefficient is not a (single) probability; it can have negative values.

There are other probabilities used in evolutionary theory that can't be interpreted in terms of actual frequencies or propensities. Coalescence theory (Hein et al. 2005) describes how you can calculate the expected number of years in the past that two or more current organisms or species have their most recent common ancestor (MRCA). This expected number takes account of several probabilities – the probability that the MRCA existed one year ago, that it existed two years ago, and so on.

[16] Recall the discussion in Section 5.8 of what it means for drift to dominate selection and the discussion in Section 8.4 of what it means for selection to exert a stronger influence than phylogenetic inertia on the evolution of a trait.

Propensities are causal powers. However, if cause must precede effect, backward-directed probabilities that have positive values don't represent propensities (Humphreys 1985).

The same conclusion applies to the $\frac{1}{p}$ argument about common ancestry discussed in Section 4.6. In that argument I discussed the value of Pr(Y has trait T | X has trait T) when X and Y are contemporaneous organisms or populations, noting that this probability has a value close to 1.0 if their most recent common ancestor existed very recently, with the probability declining if the MRCA is more ancient. This conditional probability does not describe how strong an influence X's having T exerts on whether Y has T. Your genes don't causally influence which genes your sibling has.

Should my next order of business be to find a new objective interpretation of probability that clarifies the probabilities in evolutionary theory that don't represent actual frequencies? I am disinclined to do so. The traditional project of finding "interpretations" of probability has missed a simple point. An interpretation of probability must find some antecedently well understood concept that helps one define what probability means in some or all contexts of use, but why is this type of clarification needed when we say that the probabilities used in evolutionary theory are objective? We know that probabilities must conform to the axioms of probability, and we have a grasp of their epistemology. Why do we need anything more?

The problem of giving an "interpretation" of objective probability reminds me of a problem that few philosophers now take seriously – the problem of defining the terms used in a scientific theory by using a purely observational language. This reductionist project never succeeded, and it gradually dawned on philosophers that theoretical terms often have meanings that can't be expressed in a purely observational language. Philosophers additionally saw that this failure does not represent a defect in theoretical concepts. The same anti-reductionist attitude, I suggest, makes sense for objective probabilities. The anti-reductionist idea I'm describing might be called "the no-theory theory of probability" (Sober 2010).

This skepticism about the need for a new interpretation of objective probability has the virtue of solving a problem. Philosophers have long recognized that the validity of deductive arguments depends on arguments not shifting in midstream from one meaning of a term to another. Here's an argument that shows the point of prohibiting this shift:

That building is a bank.
A bank is a side of a river.

That building is a side of a river.

Is this argument deductively valid? At first glance, it might seem to have the following logical form:

o has trait B.
For any x, if x has trait B then x has trait S.

---

o has trait S.

However, the bank argument can't be obtained from this valid argument form by *uniform substitution*, which requires that all occurrences of B are replaced by a term that has a single fixed meaning. The argument form is valid, but that doesn't mean that the bank argument is.

Now consider a derivation in evolutionary theory in which different probabilities have different interpretations. Should this derivation be scolded in the same tone of voice that philosophers use when they scold the bank argument? I don't think so. The argument consistently uses the term "probability" with a single fixed mathematical meaning, and that is enough to allow the argument to be valid.[17,18]

9.5 Contingency, Inevitability, and Sensitivity to Initial Conditions

In his book *Wonderful Life – The Burgess Shale and the Nature of History*, Stephen Jay Gould (1989) uses a vivid metaphor to pose an interesting question about the evolutionary process. The title Gould chose for his book comes from the Frank Capra movie *It's a Wonderful Life*. In that movie, the main character, George Bailey (played by Jimmy Stewart), comes to realize that his world would have been profoundly different

[17] I am not saying that the following argument is deductively valid: "$Pr_1(Y \mid X) = p$. Therefore $Pr_2(Y \mid X) = p$" where Pr_1 and Pr_2 represent different interpretations of probability (e.g., subjective credences and objective actual frequencies). Rather, I am saying that in *some* derivations, the interpretation of objective probability doesn't matter, and such derivations are routine in evolutionary theory.

[18] See Hoefer (2019) for a critique of the no-theory theory.

if he had not existed. Gould describes George's thought experiment by saying that George "replayed the tape." Gould proposes to do the same thing on a larger scale, asking what would happen if we replayed the tape of life on earth.

The homage to Capra's movie suggests that Gould's "replaying the tape" means that you imagine *changing* an event that happened earlier and then see what transpires. However, that's not how Gould describes the idea when he introduces the phrase:

> I call this experiment "replaying the tape." You press the rewind button and, making sure you thoroughly erase everything that actually happened, go back to any time and place in the past.... Then let the tape run again and see if the repetition looks anything like the original. If each replay strongly resembles life's actual pathway, then we must conclude that what happened pretty much had to occur. But suppose that the experimental versions all yield sensible results strikingly different from the actual history of life. What could we then say about the predictability of self-conscious intelligence of mammals? Or of vertebrates? Or of life on land? Or simply of multicellular persistence for 600 million difficult years? (Gould 1989, pp. 48–50)

The distinction that Gould draws here coincides pretty well with the distinction between determinism and indeterminism drawn in Section 9.2. If you start with exactly the same initial and boundary conditions, the same future must unfold again and again if the system is deterministic. However, if the system is indeterministic, the same set of initial and boundary conditions can lead to different futures.[19,20]

This interpretation of what Gould means by replaying the tape is confirmed by a hypothetical example that he then describes. Suppose that ten species survive out of a hundred, where this decimation is due to some sudden change in the environment. What will happen if you repeatedly replay the tape? Gould's answer is this:

[19] If there are five possible outcomes of replaying the tape, an indeterministic process in which they have the same probability is further from determinism than a process in which one of the outcomes has a probability of 0.99 and the other four each have a probability of 0.0025. This idea can be captured by the Shannon information measure H that Lewontin (1972) uses to discuss the biological reality of race (Section 7.8).

[20] **Question:** How would Lamarck's theory (Section 1.4) answer Gould's question about replaying the tape?

> If the ten survivors are predictable by superiority of anatomy ... then they will win each time ... but if the ten survivors are protégés of Lady Luck or fortunate beneficiaries of odd historical contingencies ..., then each replay of the tape will yield a different set of survivors and a radically different history. (p. 50)

The contrast between selection (which Gould is understanding deterministically) and Lady Luck (= pure drift) is clear, but Gould's reference to "odd historical contingencies" is puzzling. He has in mind the possibility that the ten survivors happened to inherit a trait that had no adaptive value when the trait first evolved, but suddenly proves useful when the decimation begins. It is odd that Gould thinks that this is an alternative to selection.

Gould then presents a different picture of what he means by contingency. He says that the dichotomy between determinism and pure chance is *not* what he has in mind, and that he isn't searching for a middle ground between those two extremes. The thesis he wants to defend is this:

> Alter any earlier event, ever so slightly and without apparent importance at the time, and evolution cascades into a radically different channel. This third alternative represents no more nor less than the essence of history. Its name is contingency – and contingency is a thing in itself, not a titration of determinism by randomness. (p. 51)

Gould is right that contingency (in this sense) is different from both determinism and indeterminism. Those two isms don't describe *how much* the future would be different if you changed the initial conditions (Beatty 2006b).

Gould's concept of contingency is a strong form of the idea of *sensitivity to initial conditions*, which is the defining feature of so-called *chaotic dynamics*. It is strong because he says that *all* alterations in the initial conditions (no matter how modest) would make a big difference. The phrase is often used to mean something less radical – that *some* (or many) such *changes* would have this dramatic upshot. Thanks to the mathematician and meteorologist Edward Norton Lorenz (1972), this idea is now called "the butterfly effect," wherein the occurrence of a tornado in Texas is strongly influenced by whether a butterfly in Brazil flapped its wings a few weeks earlier. If there is sensitivity to initial conditions and the laws are deterministic, a small change earlier *must* induce a large difference later; if there is sensitivity and the laws are probabilistic, a small change earlier has an *appreciable probability* of inducing a large difference later. Sensitivity to initial conditions is compatible with the existence of dynamical laws and also with both determinism and indeterminism.

Towards the end of the book, Gould characterizes contingency more cautiously:

> I am not speaking of randomness … but of a central principle of history – contingency. A historical explanation does not rely on laws of nature, but on an unpredictable sequence of antecedent states, where any *major* change in any step in the sequence would have altered the final result. (p. 283, italics mine)

The word "major" is needed here, since Gould is interested in mass extinctions; I doubt that there are many biologists who think these are caused by a single butterfly flapping its wings.[21]

Gould defends his thesis of contingency by an extended discussion of the fossils discovered in the Burgess Shale of the Canadian Rockies, which were deposited about 500 million years ago. He argues that there was much more anatomical diversity then than there is today and that most of the taxonomic groups that then existed went extinct. He also claims that it was a matter of chance which species survived – that the few survivors were no better adapted to their environments than the many that went extinct (Gould 1989, p. 236). Had the mass extinction not occurred, the subsequent history of life would have been profoundly different. Gould's idea is that this extinction event was the result of drift.

Simon Conway Morris's (2004) book, *Life's Solution – Inevitable Humans in a Lonely Universe*, is a response to Gould's argument. Replay the tape, he says, and the same broad patterns will emerge. Conway Morris is an emphatic adaptationist and this is a second disagreement he has with Gould; Conway Morris insists on the ubiquity and power of natural selection as a cause of evolutionary outcomes. The result is Darwinism without contingency. Conway Morris's term for this is "inevitability."

Conway Morris rests his argument on *convergences* (Section 7.3). Convergences have long been familiar to evolutionary biologists, but their significance, he thinks, have never been properly appreciated. The camera eye found in vertebrates has independently evolved in several other groups – squid, marine worms, jellyfish, snails, and spiders (p. 157). Conway

[21] See Swenson et al. (2000) for discussion of the butterfly effect in experiments on multi-species community selection (Section 3.8).

Morris also mentions the remarkable similarities that unite placental and marsupial mammals (see my Figure 7.1b), and he cites numerous further examples as well. He concludes that the history of life is peppered with recurrent themes; it is not an ad hoc accumulation of unique events.

Conway Morris argues that the abundance of evolutionary convergences is a decisive objection to Gould's contingency thesis. He is right if "contingency" means *maximal* sensitivity to initial conditions. We should reject the claim that the evolution of the camera eye in the vertebrate lineage depended on *all* the historical circumstances being precisely as they were. The lineage leading to present day squids accomplished the same result with different initial and boundary conditions. However, rejecting maximal contingency isn't enough to reject contingency theses that are less than maximal.

Separate problems arise for Conway Morris's positive thesis of inevitability. You can't show that the camera eye was inevitable or even highly probable just by pointing out that it evolved multiple times. To estimate the probability of the camera eye's evolving, you need to know how many times it evolved *and how many times it failed to do so.* Conway Morris never describes how often convergences *fail* to occur. The fact that the camera eye evolved n times is compatible with the eye's failing to be inevitable in each and every lineage in which the eye made an appearance. Indeed, the n successes leave open the possibility that it was very far from inevitable that the eye would evolve *at least once.*

Gould's sensitivity thesis can be reconciled with Conway Morris's inevitability thesis if Gould's concept of maximal sensitivity is replaced with something less extreme. Suppose that the n lineages in which the camera eye evolved had different initial conditions, and that each of those initial conditions made it highly probable that the camera eye would evolve. Conway Morris's inevitability thesis is true in this circumstance. However, that does not rule out the possibility that each of those initial conditions is fragile; in each lineage, there are small changes in initial conditions that would have made it very improbable that the camera eye evolves. A second reconciliation is possible. Conway Morris's thesis that convergence is rampant is compatible with Gould's thesis that it is a matter of chance which species survive mass extinctions. Imagine a series of random mass extinction events that each knocks out 90 percent of the species that then exist, where these eras of destruction are separated from each other by quieter times during which numerous convergences evolve. Perhaps Gould and Conway Morris were looking at the same rabbit/duck (Sober 2003).

Even if the two views can be reconciled in this way, the question remains of whether there can be a plausible *theory* that explains why different traits will show different degrees of sensitivity to initial conditions in their evolution. True, the camera eye evolved multiple times in lineages that differed among themselves in numerous ways. Is this a brute fact or are there deeper facts that explain why this was to be expected?

9.6 Evolution and Information Loss[22]

Is evolution an information-destroying process? The discussion of the two-state Markov process in Section 7.10 provides a clue. When a trait evolves in a lineage and the time between the process's Start (*S*) and End (*E*) is small, the likelihood ratio $\frac{\Pr(E=j \mid S=j)}{\Pr(E=j \mid S=i)}$ is much larger than 1, but as the time between *S* and *E* increases, the ratio asymptotically approaches unity. According to the law of likelihood, this means that the present state of the system strongly favors one hypothesis about the recent past over the other, but the favoring attenuates as the past becomes more and more remote.

The Markov chain convergence theorem (Cover and Thomas 2006) generalizes this pattern. Consider a system that at any time is in one of *n* possible states (s_1, s_2, ..., s_n). As before, I'll assume that the system evolves in discrete time steps. Suppose the system has the following two properties:

Irreducibility: For any two states, each is accessible from the other.

The Markov property: For any two times $t_1 < t_2$, the state of the system at t_1 screens-off the system's history prior to t_1 from the state at t_2. That is, for any i and j, Pr(system is in state j at t_2 | system is in state i at t_1) = Pr(system is in state j at t_2 | system is state i at t_1 & system's history prior to t_1).

Irreducibility does not mean that it is possible to move *directly* from every state to every other in a single jump. These two assumptions entail the following:

The Markov chain convergence theorem: If a system's evolution is irreducible and has the Markov property, then I(Past, Present) approaches zero as the time separating past and present approaches infinity.

[22] I borrow some prose in this section from Sober and Steel (2014).

I(*X*;*Y*) is the *mutual information* linking the two variables. If the variables are discrete, the formula for this quantity is:

$$I(X;Y) = \sum_{y \in Y} \sum_{x \in X} p(x,y) \log\left(\frac{p(x,y)}{p(x)p(y)}\right)$$

where $p(x,y)$ is the probability of the conjunction ($X = x$ & $Y = y$), and $p(x)$ and $p(y)$ are the probabilities of the conjuncts ($X = x$) and ($Y = y$), respectively. Mutual information measures how much (on average) you learn about the state of one variable by observing the state of the other. Its value is zero when *X* and *Y* are independent; otherwise, it is positive. Mutual information is symmetrical: I(*X*;*Y*) = I(*Y*;*X*). The decay of information described by the Markov chain convergence theorem occurs exponentially fast (Sober and Steel 2011, proposition 6). Unlike the example of the Markov process depicted in Figure 7.9, there is no requirement in the convergence theorem that transition probabilities are constant through time.

Notice that mutual information takes account of all possible states of the variables *X* and *Y* and averages over how much information observing the value of one of them would provide about the value of the other. This contrasts with the Law of Likelihood (Section 4.3), which pertains to the evidential meaning of a single observation. Another difference between mutual information and the Law of Likelihood is that mutual information is a Bayesian construct, since it requires you to consider the unconditional probabilities of hypotheses about the state of each variable. Indeed the quotient for specific values of X and Y in the definition of mutual information is equivalent to $\frac{\Pr(Y = y \mid X = x)}{\Pr(Y = y)}$; Bayesians call this the ratio measure of degree of confirmation.[23]

The convergence theorem does not ensure a monotonic decline in information as the temporal separation of past from present is increased. That extra element is provided by a different result, the so-called "data processing inequality" from information theory (Cover and Thomas 2006):

[23] Since mutual information is symmetric, the Markov chain convergence theorem entails that the present provides more information about the near future than it does about the distant future. **Question:** Is this result refuted if the probability of the First World War (W), given the entangled alliances pre-1914 (E), is greater than the probability of the war, given the assassination of Archduke Ferdinand in 1914 (A)? In answering, suppose that E caused A, and A caused W.

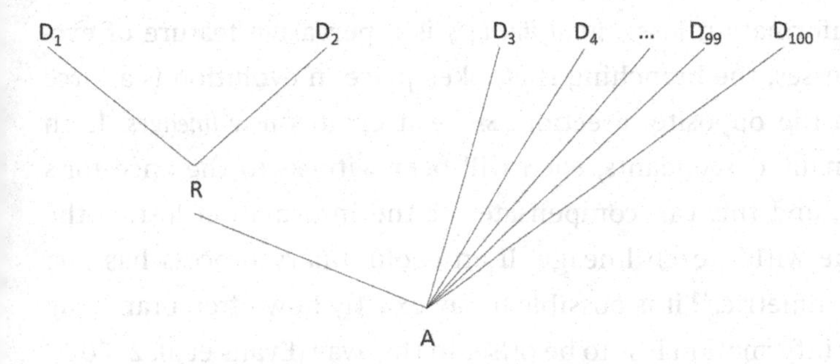

Figure 9.3 The observed states of descendants $D_1 \ldots D_{100}$ can provide more information about the state of an ancient ancestor A than they provide about the state of a recent ancestor R.

The data processing inequality: In a causal chain from a distal cause D to a proximate cause P to an effect E, if P screens-off D from E, then I(E;D) is less than or equal to both I(E;P) and I(P;D).

Here D, P, and E are variables. For a discrete-state process, these two inequalities are strict if P is probabilistically dependent on D and on E but is perfectly correlated with neither. The data processing inequality does not require the process linking D to P to be of the same kind as the process linking P to E.

The data processing inequality is "chain internal." It does not say that the present *always* provides more information about the recent past than about events that are older. To see why, consider Figure 9.3. Suppose that R screens-off A from D_1&D_2. Then the information processing inequality says that $I(D_1\&D_2; R) \geq I(D_1\&D_2; A)$. It does not say that $I(D_1\&D_2\&\ldots\&D_{100}; R) \geq I(D_1\&D_1\&\ldots\&D_{100}; A)$. It is perfectly possible that the hundred descendants provide more information about A than they provide about R. Note that R does not screen-off A from $D_1\&D_2\&\ldots\&D_{100}$.

The Markov chain convergence theorem and the data processing inequality are very general. They characterize any system whose dynamical laws have the requisite probabilistic features. The system might be an evolving population of organisms, but it could also be a volume of gas that diffuses in a room. Indeed, if there were disembodied spirits that change probabilistically, the results would apply to them. Both results are *more general than the laws of physics*. They are a priori mathematical truths, though of course it is an empirical matter whether a given system satisfies the antecedent of the conditional that each result expresses.

Although information loss *within lineages* is a pervasive feature of evolutionary processes, the branching that takes place in evolution is a force that pushes in the opposite direction, since it creates *new lineages.* If an ancestor has many descendants, each will bear witness to the ancestor's characteristics, and this can compensate for the information lost to the passage of time within each lineage. If an evolutionary process has two states and is symmetric,[24] it is possible to say exactly how often branching must occur for information loss to be offset in this way (Evans et al. 2000).[25] The mathematics is more complicated for other processes.

To get an intuitive feel for why branching can reduce information loss, consider the children's game of Telephone, in which child 1 whispers a word to child 2, who then tries to accurately repeat the word to child 3, and so on, for r rounds. The longer the chain of children, the more inaccurate you expect the last child's report about the first child's utterance to be. There is no branching in this game, but it is easy to see what branching would involve. Suppose child 1 whispers the word to n children, each of whom whispers the word they think they heard to n children and so on, for r rounds, at which time there are n^r children at the leaves of a branching tree. The reports of the n^r children in the final round of the game can be more informative than the reports of the last child in the usual chain-version of Telephone.

Just as the fact of branching has an impact on information loss, so too does the topology of the branching process itself. Consider the following conjecture about the two genealogies depicted in Figure 9.4: Observing the four descendants will provide more information about their most recent common ancestor if they are related to each other by a star phylogeny than would be the case if they were related to each other by a balanced bifurcating tree. The conjecture seems reasonable, since in Figure 9.4a, the observations are all probabilistically independent of each other, conditional on the state of the root, whereas in Figure 9.4b, the observations aren't all

[24] As noted in Section 4.9, a process is symmetric if and only if Pr(system is in state j at t_2 | system is in state i at t_1) = Pr(system is in state i at t_2 | system is in state j at t_1), for all i, j. Drift is a symmetric process, but selection for state i and against state j is not.

[25] **Question:** Is information loss when you look within a lineage and a failure of information loss when you look across lineages an instance of Simpson's paradox (Section 3.2)?

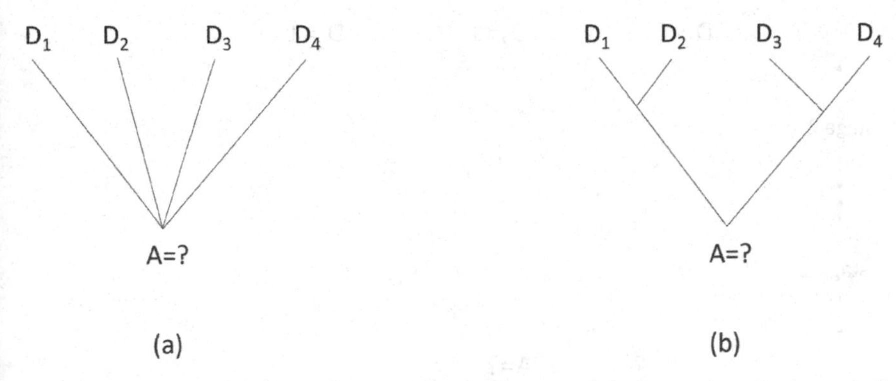

Figure 9.4 The topology of a tree affects how much information the states of descendants provide about the state of a common ancestor.

conditionally independent.[26] To focus on a simple example, let's assume cross-branch and cross-model homogeneity (Section 4.4) – that the same process is at work in all the branches in the two genealogies. The conjecture holds when there is a symmetric two-state process at work in branches (Evans et al. 2000), but it sometimes fails in symmetric processes with more than two states (Sly 2011).

To apply the concept of mutual information to the problem depicted in Figure 9.4, you need to take account of all the possible states that an ancestor and its descendants might occupy. Now let's shift to the different question depicted in Figure 9.5. Given the assumptions discussed in Section 4.4, the observation that descendants D_1 and D_3 are in state 1 favors the hypothesis that ancestor A was in state 1 over the hypothesis that A was in state 0, and the degree of that favoring is represented by the likelihood ratio $\frac{\Pr(D_1 = 1 \,\&D_3 = 1 \mid A = 1)}{\Pr(D_1 = 1 \,\&D_3 = 1 \mid A = 0)}$. Similarly, the observation that descendants D_1 and D_2 are in state 1 favors the hypothesis that ancestor A was in state 1 over the hypothesis that A was in state 0, and the degree of that favoring is represented by the likelihood ratio $\frac{\Pr(D_1 = 1 \,\&D_2 = 1 \mid A = 1)}{\Pr(D_1 = 1 \,\&D_2 = 1 \mid A = 0)}$. The question now is whether the first likelihood ratio is bigger than the second. A plausible guess is that the first ratio is greater than the second because the

[26] Notice that I'm using the concept of conditional independence here, not the concept of unconditional independence (Section 1.1). Question: Do the two topologies in Figure 9.4 differ with respect to whether the leaves are unconditionally dependent on each other?

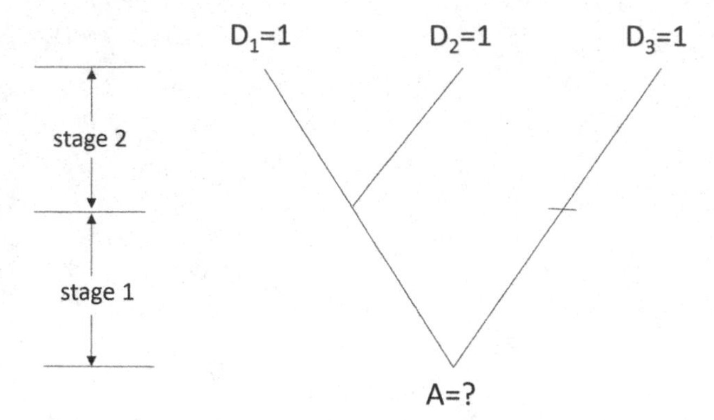

Figure 9.5 Comparing independent and dependent evidence about the state of a common ancestor.

states of D_1 and D_3 are probabilistically independent of each other, conditional on the state of ancestor A, whereas the states of D_1 and D_2 are not independent of each other, again conditional on the state of A.[27] The idea that independent evidence is better is suggested by a wry example that Wittgenstein (1953, §265) describes in *Philosophical Investigations*: a man is skeptical about a story he reads in the newspaper, so he buys an additional copy of the same newspaper to double-check (Sober 1989).

To analyze the evidential relevance of the distinction between conditionally independent and conditionally dependent observations depicted in Figure 9.5, let's examine a simple model in which all the branches in this two-stage process obey the same transition probabilities, where s = Pr(branch ends in state 1 | branch starts in state 1) and e = Pr(branch ends in state 1 | branch starts in state 0). The values of s and e are constrained to be such that $s > e$; this is the COR inequality (that descendants are positively correlated with their ancestors) discussed in Chapter 4. Figure 9.6 describes when the conjunctive observation ($D_1 = 1$ & $D_3 = 1$) favors $A = 1$ over $A = 0$ more strongly than the conjunctive observation ($D_1 = 1$ & $D_2 = 1$) does, and when the reverse is true.[28] As you see in the

[27] The difference between mutual information and the law of likelihood is an instance of Hacking's (1965, pp. 95–102) distinction between before-test betting and after-test evaluation.

[28] My thanks to William Roche for finding counterexamples to the thesis that conditionally independent evidence is always stronger and to Matthew Maxwell for preparing Figure 9.6.

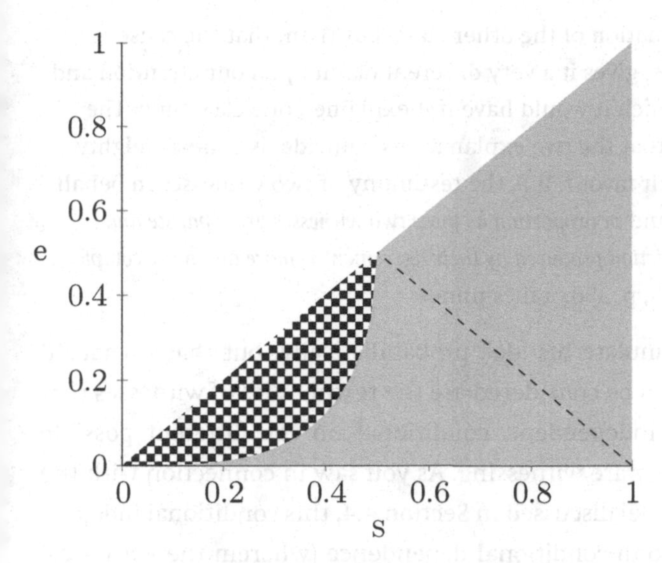

Figure 9.6 An analysis of the problem posed in Figure 9.5. In the solid gray region, the conjunctive observation ($D_1 = 1$ & $D_3 = 1$) favors A = 1 over A = 0 more strongly than the conjunctive observation ($D_1 = 1$ & $D_2 = 1$) does. The reverse is true in the square-hatched region. The broken line represents the symmetrical case in which $s = 1 - e$.

figure, there are exceptions to the maxim that independent evidence is always stronger. It is interesting that counterexamples exist only when $s < 0.5$. In terms of the Markov model described in Chapter 7, this situation arises only when there was selection for state 0 and enough time in branches to make evolution from 1 to 0 more probable than the retention of state 1.[29]

The idea that independent sources of evidence are better than dependent sources has a history. Philosophers often cite William Whewell's *consilience of inductions* in this regard, which he describes as

> one of the most important tests which can be given of a sound physical theory. It is true, the explanation of one set of facts may be of the same

[29] **Exercise:** The simple model used here to show that independent evidence isn't always stronger than dependent is unrealistic. Does that undermine the force of the argument presented here? Compare this example with the disagreement between Felsenstein and Farris concerning the statistical inconsistency of cladistic parsimony (Section 7.12).

> nature as the explanation of the other class: but then, that the cause explains *both* classes, gives it a very different claim upon our attention and assent from that which it would have if it explained one class only. The very circumstance that the two explanations coincide, is a most weighty presumption in their favour. It is the testimony of two witnesses in behalf of the hypothesis; and *in proportion as these two witnesses are separate and independent, the conviction produced by their agreement is more and more complete.* (Whewell 1840/1968, p. 330; italics mine)

Whewell doesn't formulate his idea probabilistically, but that's what I'll do now. What needs to be considered are the testimonies of witnesses that are probabilistically independent, conditional on the different possible states of the event they are witnessing. As you saw in connection with the common ancestry model discussed in Section 4.4, this conditional independence can give rise to unconditional dependence (wherein the witnesses' testimonies are correlated). The upshot is the agreement among testimonies that Whewell describes.

I mentioned in Chapter 6 that Whewell was Darwin's teacher at the University of Cambridge. Whewell never accepted Darwin's theory, even though Darwin thought he was using Whewell's epistemology (Waters 2003). Whewell's influence is evident in the following passage from Darwin's *Variation of Animals and Plants under Domestication*:

> In scientific investigations it is permitted to invent any hypothesis, and if it explains *various large and independent classes of facts* it rises to the rank of a well-grounded theory. The undulations of the ether and even its existence are hypothetical, yet every one now admits the undulatory theory of light. The principle of natural selection may be looked at as a mere hypothesis, but rendered in some degree probable by what we already know of the variability of organic beings in a state of nature, – by what we positively know of the struggle for existence, and the consequent almost inevitable preservation of favourable variations, – and from the analogical formation of domestic races. Now this hypothesis may be tested, – and this seems to me the only fair and legitimate manner of considering the whole question, – *by trying whether it explains several large and independent classes of facts*; such as the geological succession in organic beings, their distribution in past and present times, and their mutual affinities and homologies. If the principle of natural selection does explain these and other large bodies of facts, it ought to be received (Darwin 1868, vol. 1, p. 9; italics mine).

It is ironic that an example involving phylogenetic inference should pose a problem for the Whewellian principle that influenced Darwin. What is called for, however, is not a wholesale rejection of the claim that independent evidence is better than dependent. The way forward is to find more modest and conditional formulations.

References

Abrams, M. (2007). "How Do Natural Selection and Random Drift Interact?" *Philosophy of Science* 74(5): 666–679.

Akaike, H. (1973). "Information Theory and an Extension of the Maximum Likelihood Principle." In B. Petrov and F. Csaki (eds.), *Second International Symposium on Information Theory*. Akademiai Kiado, pp. 267–281.

Andreasen, R. (1998). "A New Perspective on the Race Debate." *British Journal for the Philosophy of Science* 49: 199–225.

Axelrod, R. (1984). *The Evolution of Cooperation*. Basic Books.

Barrett, M., Clatterbuck, H., Goldsby, M. et al. (2012). "Puzzles for ZFEL – McShea and Brandon's Zero Force Evolutionary Law." *Biology and Philosophy* 27(5): 723–735.

Baum, D. (1998). "Individuality and the Existence of Species through Time." *Systematic Biology* 47(4): 641–653.

Baum, D. (2008). "Reading a Phylogenetic Tree – The Meaning of Monophyletic Groups." *Nature Education* 1(1): 190.

Baum, D. (2009). "Species as Ranked Taxa." *Systematic Biology* 58(1): 74–86.

Baum, D. and Smith, S. (2013). *Tree Thinking – An Introduction to Phylogenetic Biology*. Roberts & Co.

Beatty, J. (1984). "Chance and Natural Selection." *Philosophy of Science* 51(2): 183–211.

Beatty, J. (1992). "Random Drift." In E. Keller and E. Lloyd (eds.), *Keywords in Evolutionary Biology*. Harvard University Press, pp. 273–281.

Beatty, J. (1995). "The Evolutionary Contingency Thesis." In G. Wolters and J. Lennox (eds.), *Concepts, Theories, and Rationality in the Biological Sciences*. University of Pittsburgh Press, pp. 45–81.

Beatty, J. (2006a). "Chance Variation – Darwin on Orchids." *Philosophy of Science* 73(5): 629–641.

Beatty, J. (2006b). "Replaying Life's Tape." *Journal of Philosophy* 103(7): 336–362.

Beatty, J. and Finsen, S. (1989). "Rethinking the Propensity Interpretation: A Peek Inside Pandora's Box." In M. Ruse (ed.), *What the Philosophy of Biology Is*. Kluwer, pp. 17–30.

Bersaglieri, T., Sabeti, P. C., Patterson, N. et al. (2004). "Genetic Signatures of Strong Recent Positive Selection at the Lactase Gene." *American Journal of Human Genetics* 74(6): 1111–1120.

Bethell, T. (1976). "Darwin's Mistake." *Harper's* 252(1509): 70

Bier, E. and Sober, E. (2020). "CRISPR Gene Drive and the War against Malaria." *American Scientist* 108(3): 162–169.

Blomberg, S., Garland, T., and Ives, A. (2003). "Testing for Phylogenetic Signal in Comparative Data – Behavioral Traits Are More Labile." *Evolution* 57(4): 717–745.

Bois, J. (2016). "The Luria– Delbrück Fluctuation Experiment." http://justinbois.github.io/bootcamp/2016/lessons/l29_luria_delbruck.html.

Bourrat, P. (2021). *Facts, Conventions, and Levels of Selection.* Cambridge University Press.

Boyd, R. (1999). "Homeostasis, Species, and Higher Taxa." In R. Wilson (ed.), *Species – New Interdisciplinary Essays.* MIT Press, pp. 141–185.

Boyd, R. and Richerson, P. (1985). *Culture and the Evolutionary Process.* University of Chicago Press.

Bradley, B. (2021). *Darwin's Psychology.* Oxford University Press.

Brandon, R. (1978). "Adaptation and Evolutionary Theory." *Studies in the History and Philosophy of Science* 9(3): 181–206.

Brandon, R. (2005). "The Difference between Selection and Drift: A Reply to Millstein." *Biology and Philosophy* 20(1): 153–170.

Brandon, R. and Fleming, L. (2014). "Drift Sometimes Dominates Selection, and Vice Versa – A Reply to Clatterbuck, Sober and Lewontin." *Biology and Philosophy* 29: 577–585.

Bridgman, P. (1927). *The Logic of Modern Physics.* Macmillan.

Brigandt, I. and Love, A. (2017). "Reductionism in Biology." In E. Zalta (ed.), *Stanford Encyclopedia of Philosophy.* https://plato.stanford.edu/archives/sum2023/entries/reduction-biology/

Buss, L. (1987). *The Evolution of Individuality.* Princeton University Press.

Byerly, H. and Michod, R. (1991). "Fitness and Evolutionary Explanation." *Biology and Philosophy* 6(1): 45–53.

Carnap, R. (1950). *Logical Foundations of Probability.* University of Chicago Press.

Cartwright, N. (1979). "Causal Laws and Effective Strategies." *Noûs* 13(4): 419–437.

Cavalli-Sforza, L. (1991). "Genes, Peoples and Languages." *Scientific American* 265(5): 104–110.

Cavalli-Sforza, L. and Feldman, M. (1981). *Cultural Transmission and Evolution: A Quantitative Approach.* Princeton University Press.

Chang, H. (2021). "Operationalism." In E. Zalta (ed.), *The Stanford Encyclopedia of Philosophy.* https://plato.stanford.edu/archives/fall2021/entries/operationalism/

Choi, S. and Fara, M. (2021) "Dispositions." In E. Zalta (ed.), *The Stanford Encyclopedia of Philosophy*. https://plato.stanford.edu/archives/spr2021/entries/dispositions/

Clarke, B. (1971). "Natural Selection and the Evolution of Proteins." *Nature* 232(5311): 487.

Clatterbuck, H. (2022). "Darwin's Causal Argument against Intelligent Design." *Philosopher's Imprint*. DOI: https://doi.org/10.3998/phimp.930.

Clatterbuck, H., Sober, E., and Lewontin, R. (2013). "Selection Never Dominates Drift (nor Vice Versa)." *Biology and Philosophy* 28(4): 577–592.

Coddington, J. (1988). "Cladistic Tests of Adaptational Hypotheses." *Cladistics* 4(1): 3–22.

Conway Morris, S. (2004). *Life's Solution – Inevitable Humans in a Lonely Universe*. Cambridge University Press.

Cover, T. M. and Thomas, J. A. (2006). *Elements of Information Theory*, 2nd edition. Wiley.

Coyne, J. and Orr, H. A. (2004). *Speciation*. Oxford University Press.

Crick, F. H. C. (1957). "Nucleic Acids." *Scientific American* 197(3): 188–203.

Cummins, R. (1975). "Functional Analysis." *Journal of Philosophy* (72): 741–764.

Cunningham, D., Omland, K., and Oakley, T. (1998). "Reconstructing Ancestral Character States – A Critical Reappraisal." *Trends in Ecology and Evolution* 13(9): 361–366.

Cutter, A. D. (2019). *A Primer of Molecular Population Genetics*. Oxford University Press.

Damuth, J. and Heisler, I. (1988). "Alternative Formulations of Multilevel Selection." *Biology and Philosophy* 3(4): 407–430.

Darwin, C. (1859). *On the Origin of Species by Means of Natural Selection*. Murray. Harvard University Press, 1964.

Darwin, C. (1860). "Letter of Asa Gray." Letter no. 2998 in the Darwin Correspondence Project. www.darwinproject.ac.uk/letter/?docId=letters/DCP- LETT- 2998.xml.

Darwin, C. (1860). "Letter to J.D. Hooker." Letter no. 2816 in the Darwin Correspondence Project. www.darwinproject.ac.uk/letter/?docId=letters/DCP- LETT- 2816.xml.

Darwin, C. (1862/1867). *On the Various Contrivances by Which British and Foreign Orchids Are Fertilised by Insects*. Murray. References to the revised edition of 1867, reprinted by University of Chicago Press, 1984.

Darwin, C. (1868). *The Variation in Animals and Plants under Domestication*. Murray.

Darwin C. (1871). *The Descent of Man and Selection in Relation to Sex*. Murray.

Darwin, C. (1958). *The Autobiography of Charles Darwin 1809–1882 with the Original Omissions Restored*. N. Barlow (ed.), Collins.

Darwin, C. (1959): *On the Origin of Species – A Variorum Edition*. M. Peckham (ed.), University of Pennsylvania Press.

Dawkins, R. (1976). *The Selfish Gene*. Oxford University Press.

Dawkins, R. (1982). *The Extended Phenotype – The Gene as the Unit of Selection*. Oxford University Press.

De Queiroz, K. (2005). "Different Species Problems and Their Resolution." *BioEssays* 27(120): 1263–1269.

Devitt, M. (2008). "Resurrecting Biological Essentialism." *Philosophy of Science* 75(3): 344–382.

Diaconis, P. (1998). "A Place for Philosophy? The Rise of Modeling in Statistical Science." *Quarterly of Applied Mathematics* 56(4): 797–805.

Dietrich, M. (1994). "The Origins of the Neutral Theory of Molecular Evolution." *Journal of the History of Biology* 27(1): 21–59.

Dietrich, M. and Millstein, R. (2008). "The Role of Causal Processes in the Neutral and Nearly Neutral Theories." *Philosophy of Science* 75(5): 548–559.

Diez, J. and Lorenzano, P. (2015). "Are Natural Selection Explanatory Models *A Priori*?" *Biology and Philosophy* 30(6): 787–809.

Dixon, T. M. (2008). *The Invention of Altruism – Making Moral Meanings in Victorian Britain*. Oxford University Press.

Dobzhansky, T. and Pavlovsky, O. (1957). "An Experimental Study of Interaction between Genetic Drift and Natural Selection." *Evolution* 11(3): 311–319.

Doolittle, W. F. (2000). "Uprooting the Tree of Life." *Scientific American* 282(2): 90–95.

Doolittle, W. F. and Inkpen, S. A. (2018). "Processes and Patterns of Interaction as Units of Selection: An Introduction to ITSNTS Thinking." *Proceedings of the National Academies of Science USA* 115(16): 4006–4014.

Doudna, J. and Charpentier, R. (2014). "The New Frontier of Genome Engineering with CRISPR- Cas9." *Science* 346(6213): 1077.

Dretske, F. (1988). *Explaining Behavior – Reasons in a World of Causes*. MIT Press.

Dugatkin, L. A. and Reeve, H. K. (1994). "Behavioral Ecology and Levels of Selection – Dissolving the Group Selection Controversy." *Advances in the Study of Behavior* 23: 101–133.

Duhem, P. (1914) "La Théorie Physique – Son Objet, Sa Structure. Chevalier and Riviére." 2nd edition. English translation as *The Aim and Structure of Physical Theory*, Princeton University Press, 1954.

Dunn, L. C. (1953). "Variations in the Segregation Ratio as Causes of Variations of Gene Frequency." *Acta Genetica et Statististica Medica* 4(2–3): 139–147.

Dupré, J. (2012). *Processes of Life – Essays in Philosophy of Biology*. Oxford University Press.

Earman, J. and Friedman, M. (1973). "The Meaning and Status of Newton's Law of Inertia and the Nature of Gravitational Forces." *Philosophy of Science* 40(3): 329–359

Earman, J. (1986). *A Primer on Determinism*. D. Reidel.

Edwards, A. (1972). *Likelihood*. Cambridge University Press.

Edwards, A. (2000). "Carl Düsing (1884) on the Regulation of the Sex–Ratio." *Theoretical Population Biology* 58(3): 255–257.

Edwards, A. (2003). "Human Genetic Diversity – Lewontin's Fallacy." *BioEssays* 25(8): 798–801.

Edwards, A. (2007). "Maximisation Principles in Evolutionary Biology." In M. Matthen and C. Stephens (eds.), *Philosophy of Biology*. North Holland, pp. 347–359.

Eells, E. (1991). *Probabilistic Causality*. Cambridge University Press.

Eldredge, N. and Cracraft, J. (1980). *Phylogenetic Patterns and the Evolutionary Process*. Columbia University Press.

Eldredge, N. and Gould, S. (1972). "Punctuated Equilibria, an Alternative to Phyletic Gradualism." In T. Schopf (ed.), *Models in Paleobiology*. Freeman Cooper, pp. 82–115.

Elgin, M. (2003). "Biology and *A Priori* Laws." *Philosophy of Science* 70(5): 1380–1389.

Elgin, M. and Sober, E. (2015). "Causal, *A Priori* True, and Explanatory – A Reply to Lange and Rosenberg." *Australasian Journal of Philosophy* 93(1): 167–171.

Elgin, M. and Sober, E. (2017). "Popper's Shifting Appraisal of Evolutionary Theory." *Hopos – Journal of the International Society for the History of Philosophy of Science* 7(1): 31–55.

Ereshefsky, M. (2010). "Microbiology and the Species Problem." *Biology and Philosophy* 25: 553–568.

Ereshefsky, M. (2014). "Species, Historicity, and Path Dependency." *Philosophy of Science* 81(5): 714–726.

Ereshefsky, M. (2022). "Species." In E. Zalta (ed.), *The Stanford Encyclopedia of Philosophy*. https://plato.stanford.edu/archives/sum2022/entries/species/

Evans, W., Kenyon, C., Peres, Y., and Schulman, L. (2000). "Broadcasting on Trees and the Ising Model." *Advances in Applied Probability* 10(2): 410–433.

Farris, J. S. (1983). "The Logical Basis of Phylogenetic Analysis." In N. Platnick and V. Funk (eds.), *Advances in Cladistics – Proceedings of the 2nd Annual Meeting of the Willi Hennig Society*. Columbia University Press, pp. 7–36.

Feldman, M. and Lewontin, R. (2008). "Race, Ancestry, and Medicine." In B. Koenig, S. Lee, and S. Richardson (eds.), *Revisiting Race in a Genomic Age*. Rutgers University Press, pp. 89–101.

Felsenstein, J. (1973). "Maximum Likelihood and Minimum – Step Methods for Estimating Evolutionary Trees from Data on Discrete Characters." *Systematic Zoology* 22(3): 240–249.

Felsenstein, J. (1978). "Cases in Which Parsimony and Compatibility Methods Can Be Positively Misleading." *Systematic Biology* 27(3): 401–410.

Fisher, R. (1922). "On the Mathematical Foundations of Theoretical Statistics." *Philosophical Transactions of the Royal Society* 222(594–604): 309–368.

Fisher, R. (1925). *Statistical Methods for Research Workers.* Oliver and Boyd.

Fisher, R. (1930). *The Genetical Theory of Natural Selection.* Clarendon Press.

Fisher, R. (1938). "Comment on H. Jeffrey's 'Maximum Likelihood, Inverse Probability, and the Method of Moments.'" *Annals of Eugenics* 8(2): 146–151.

Fisher, R. (1950). *Statistical Methods for Research Workers.* Oliver and Boyd.

Fisher, R. A. (1956). *Statistical Methods and Scientific Inference.* Hassell Street Press.

Fitelson, B. (2011). "Favoring, Likelihoodism, and Bayesianism." *Philosophy and Phenomenological Research* 83(3): 666–672.

Fitelson, B. and Hitchcock, C. (2011). "Probabilistic Measures of Causal Strength." In P. McKay Illari and F. Russo (eds.), *Causality in the Sciences.* Oxford University Press, pp. 600–627.

Fitzgerald, D. and Rosenberg, S. (2019). "What Is Mutation? A Chapter in the Series: How Microbes 'Jeopardize' the Modern Synthesis." *PLoS Genetics* 15(4): e1007995.

Fleeming Jenkin, H. (1867). "Review of *The Origin of Species.*" *North British Review* 46(92): 277–318.

Fodor, J. (1974). "Special Sciences." *Synthese* 28(2): 97–115.

Forster, M. and Sober, E. (1994). "How to Tell When Simpler, More Unified, or Less *Ad Hoc* Theories Will Provide More Accurate Predictions." *The British Journal for the Philosophy of Science* 45(1): 1–35.

Frank, S. and Slatkin, M. (1990). "Evolution in a Variable Environment." *American Naturalist* 136(2): 244–260.

Freeland, S., Knight, R., Landweber, L., and Hurst, L. (2000). "Early Fixation of an Optimal Genetic Code." *Molecular Biology and Evolution* 17(4): 511–518.

Galavotti, M. C. (2015). *A Philosophical Introduction to Probability.* Center for the Study of Language and Information.

Galton, F. (1869). *Hereditary Genius.* Appleton.

Galton, F. (1883). *Inquiries into Human Faculty and Its Development.* Macmillan.

Garson, J. (2023). *What Biological Functions Are and Why They Matter.* University of Cambridge Press.

Gaut, B. and Lewis, P. (1995). "Success of Maximum Likelihood Phylogeny Inference in the Four Taxon Case." *Molecular Biology and Evolution* 12: 152–162.

Geoghegan, J. and Holmes, E. (2018). “The Phylogenomics of Evolving Virus Virulence,” *Nature Reviews – Genetics* 19: 756–769.

Gerbault, P., Liebert, A., Itan, Y. et al. (2011). “Evolution of Lactase Persistence: an Example of Human Niche Construction.” *Philosophical Transactions of the Royal Society B: Biological Sciences* 366(1566): 863–877.

Ghiselin, M. (1974). “A Radical Solution to the Species Problem.” *Systematic Zoology* 23(4): 536–544.

Gigerenzer, G., Swijtink, Z., Porter, T. et al. (1989). “Chance and Life – Controversies in Modern Biology.” In, *The Empire of Chance – How Probability Changed Science and Everyday Life*. Cambridge University Press, pp. 132–162.

Gilbert, S., Sapp, J., and Tauber, A. (2012). “Symbiotic View of Life – We Have Never Been Individuals.” *Quarterly Review of Biology* 87: 325–341.

Gildenhuys, P. (2003). “The Evolution of Altruism – The Sober/Wilson Model.” *Philosophy of Science* 70(1): 27–48.

Gildenhuys, P. (2009). “An Explication of the Causal Dimension of Drift” *British Journal for the Philosophy of Science* 60(3): 521–555.

Gillespie, J. H. (1974). “Natural Selection for Within – Generation Variance in Offspring Number.” *Genetics* 76(3): 601–606.

Gillies, D. (2016). “The Propensity Interpretation.” In A. Hájek and C. Hitchcock (eds.), *The Oxford Handbook of Probability and Philosophy*. Oxford University Press, pp. 406–422.

Glasgow, J., Haslanger, S., Jeffers, C., and Spencer, Q. (2019). *What Is Race? Four Philosophical Views*. Oxford University Press.

Glennan, S. (1997). “Probable Causes and the Distinction between Subjective and Objective Chance.” *Nous* 4: 496–518.

Glymour, B. (2001). “Selection, Indeterminism, and Evolutionary Theory.” *Philosophy of Science* 68: 518–535.

Glymour, C., Spirtes, P., and Scheines, R. (1993). *Causation, Prediction and Search*. Springer, 2nd edition. MIT Press, 2001.

Goodman, N. (1965). *Fact, Fiction and Forecast*, 2nd edition. Bobbs–Merrill.

Godfrey-Smith, P. (2001). “Three Kinds of Adaptationism.” In S. Orzack and E. Sober (eds.), *Adaptationism and Optimality*. Cambridge University Press, pp. 303–334.

Godfrey-Smith, P. (2012). “Darwinism and Cultural Change.” *Philosophical Transactions of the Royal Society B* 367: 2160–2170.

Goodnight, C. (1990). “Experimental Studies in Community Evolution 1 – The Response to Selection at the Community Level.” *Evolution* 44(6): 1614–1624.

Gottlieb, P. and Sober, E. (2017). “Aristotle on ‘Nature does Nothing in Vain.’” *Hopos – The Journal of the International Society for the History of Philosophy of Science* 7(2): 246–271.

Gould, S. (1977a). "Darwin's Untimely Burial." In *Ever Science Darwin*. W. W. Norton & Company, pp. 39–48.

Gould, S. (1977b). "The Misnamed, Mistreated, and Misunderstood Irish Elk." In *Ever Since Darwin*. W. W. Norton & Company, pp. 79–90.

Gould, S. (1989). *Wonderful Life – The Burgess Shale and the Nature of History*. W. W. Norton & Company.

Gould, S. and Lewontin, R. (1979). "The Spandrels of San Marco – A Critique of the Adaptationist Programme." *Proceedings of the Royal Society of London B* 205(1161): 581–598.

Grafen, A. (1991). "Modeling in Behavioral Ecology." In J. Krebs and N. Davies (eds.), *Behavioral Ecology – An Evolutionary Approach*. Blackwell Scientific Publications, pp. 5–31.

Graves, L., Horan, B. and Rosenberg, A. (1999). "Is Indeterminism the Source of the Statistical Character of Evolutionary Theory?" *Philosophy of Science* 66: 140–157.

Gray, A. (1860). "Darwin on the Origin of Species." *Annals and Magazine of Natural History* 6(35): 373–386. Reprinted in *Darwiniana*, Harvard University Press, 1963, pp. 72–45.

Griffiths, P. (1996). "The Historical Turn in the Study of Adaption." *British Journal for the Philosophy of Science* 47(4): 511–532.

Griffiths, P. (1999). "Squaring the Circle – Natural Kinds with Historical Essences." In R. A. Wilson (ed.), *Species: New Interdisciplinary Essays*. MIT Press, pp. 208–228.

Griffiths, P. (2020). "Sex Is Real." *Aeon*. https://aeon.co/essays/the-existence-of-biological-sex-is-no-constraint-on-human-diversity

Grimmett, G. and Stirzaker, D. (2001). *Probability and Random Processes*, 3rd edition. Oxford University Press.

Haber, M. and Velasco, J. (2022). "Phylogenetic Inference." In E. Zalta (ed.), *The Stanford Encyclopedia of Philosophy*. https://plato.stanford.edu/archives/fall2022/entries/phylogenetic-inference/

Hacking, I. (1965). *The Logic of Statistical Inference*. Cambridge University Press.

Hájek, A. (2019). "Interpretations of Probability." In E. Zalta (ed.), *The Stanford Encyclopedia of Philosophy*. https://plato.stanford.edu/archives/fall2019/entries/probability-interpret/

Hamilton, W. D. (1964). "The Genetical Evolution of Social Behaviour I and II." *Journal of Theoretical Biology* 7(1): 1–16, 17–52.

Hamilton, W. D. (1967). "Extraordinary Sex Ratios." *Science* 156(3774): 477–488.

Hamilton, W. D. (1971). "Geometry for the Selfish Herd." *Journal of Theoretical Biology* 31(2): 295–311.

Hansen, T. F. (2017). "On the Definition and Measurement of Fitness in Finite Populations." *Journal of Theoretical Biology* 419: 36–43.

Hartl, D. and Clark, A. (2007). *Principles of Population Genetics*, 4th edition. Sinauer Associates.

Hausman, D. M. and Woodward, J. (1999). "Independence, Invariance and the Causal Markov Condition." *The British Journal for the Philosophy of Science* 50(4): 521–583.

Heath, J. (2020). "Methodological Individualism." In E. Zalta (ed.), *Stanford Encyclopedia of Philosophy*. https://plato.stanford.edu/archives/sum2020/entries/methodological-individualism/

Hein, J, Schierup, M., and Wiuf, C. (2005). *Gene Genealogies, Variation and Evolution – A Primer in Coalescent Theory*. Oxford University Press.

Heisler, I. L. and Damuth, J. (1987). "A Method for Analyzing Selection in Hierarchically Structured Populations." *American Naturalist* 130(4): 582–602.

Helgeson, C. (2018). "Modus Darwin Reconsidered." *British Journal for the Philosophy of Science* 69(1): 193–213.

Hempel, C. (1965). "Aspects of Scientific Explanation." In *Aspects of Scientific Explanation and Other Essays in Philosophy of Science*. Free Press, pp. 331–497.

Hennig, W. (1950/1966). *Grundzüge einer Theorie der phylogenetischen Systematik*, Deutscher Zentralverlag. Translated by D. Davis and R. Zangerl as *Phylogenetic Systematics*, University of Illinois Press.

Herbert, S. (1971). "Darwin, Malthus, and Selection." *Journal of the History of Biology* 30(4): 209–217.

Hillis, D., Huelsenbeck, J., and Cunningham, C. (1993). "Application and Accuracy of Molecular Phylogenies." *Science* 264: 671–676.

Hitchcock, C. and Redei, M. (2021). "Reichenbach's Common Cause Principle." In E. Zalta (ed.), *Stanford Encyclopedia of Philosophy*. https://plato.stanford.edu/archives/sum2021/entries/physics-Rpcc/

Hodge, J. (1987). "Natural Selection as a Causal, Empirical, and Probabilistic Theory." In L. Kruger, G. Gigerenzer, and M. Morgan (eds.), *The Probabilistic Revolution*, vol. 2. MIT Press, pp. 233–270.

Hoefer, C. (2019). *Chance in the World – A Humean Guide to Objective Chance*. Oxford University Press.

Hrdy, S. (1997). "Raising Darwin's Consciousness." *Human Nature* 8(1): 1–49.

Hubby, J. and Lewontin, R. (1966). "A Molecular Approach to the Study of Genic Heterozygosity in Natural Populations. 1. The Number of Alleles at Different Loci in *Drosophila pseudoobscura*." *Genetics* 54(2): 577–594.

Hull, D. (1965). "The Effect of Essentialism on Taxonomy – Two Thousand Years of Stasis." *British Journal for Philosophy of Science* 15(60): 314–326, 16(61): 1–18.

Hull, D. (1968). "The Operational Imperative – Sense and Nonsense in Operationalism." *Systematic Zoology* 17(4): 438–457.

Hull, D. (1970). "Contemporary Systematic Philosophies." *Annual Review of Ecology and Systematics* 1: 19–54.

Hull, D. (1978). "A Matter of Individuality." *Philosophy of Science* 45(3): 335–360.

Hull, D. (1988). *Science as a Process.* University of Chicago Press.

Humphreys, P. (1985). "Why Propensities Cannot be Probabilities." *Philosophical Review* (94)4: 557–570.

Huxley, T. H. (1860). "The Origin of Species." *Westminster Review* 17: 541–570.

Jablonka, E. and Lamb, M. (2005). *Evolution in Four Dimensions – Genetic, Epigenetic, Behavioral, and Symbolic Variation in the History of Life*. MIT Press.

James, M. and Burgos, A. (2022). "Race." In E. Zalta (ed.), *The Stanford Encyclopedia of Philosophy.*

Jenkins, F. (1867). "The Origin of Species." *North British Review* 46: 277–318.

Jensen A. (1969). "How Much Can We Boost IQ and Scholastic Achievement?" *Harvard Education Review* 39(1): 1–123.

Johnson, L. (1970). "Rainbow's End – The Quest for an Optimal Taxonomy." *Systematic Zoology* 19(3): 203–239.

Keller, J. (1986). "The Probability of Heads." *American Mathematical Monthly* 93(3): 191–197.

Kevles, D. (1998). *In the Name of Eugenics – Genetics and the Uses of Human Heredity.* Harvard University Press.

Kimura, M. (1957). "Some Problems of Stochastic Processes in Genetics." *Annals of Mathematical Statistics* 28(4): 882–901.

Kimura, M. (1962). "On the Probability of Fixation of Mutant Genes in a Population." *Genetics* 47(6): 713–719.

Kimura, M. (1967). "On the Evolutionary Adjustment of Spontaneous Mutation Rates." *Genetics Research* 9(1): 23–34.

Kimura, M. (1968). "Evolutionary Rate at the Molecular Level." *Nature* 217(5129): 624–626.

Kimura, M. (1983). *The Neutral Theory of Molecular Evolution.* Cambridge University Press.

Kimura, M. (1991). "The Neutral Theory of Molecular Evolution – A Review of Recent Evidence." *Japanese Journal of Genetics* 66(4): 367–386.

Kimura, M. and Ohta, T. (1971). "Protein Polymorphism as a Phase of Molecular Evolution." *Nature* 229(5285): 467–469.

King, J. and Jukes, T. (1969). "Non- Darwinian Evolution." *Science* 164(3881): 788–798.

Kishino, H. and Hasegawa, M. (1989). "Evaluation of the Maximum Likelihood Estimate of the Evolutionary Tree Topologies from DNA Sequence Data, and the Branching Order in Hominoidea." *Journal of Molecular Evolution* 29(2): 170–179.

Kitcher, P. (1990). "The Division of Cognitive Labor." *Journal of Philosophy* 87(1): 5–22.

Kitcher, P., Sterelny, K., and Waters, K. (1990). "The Illusory Riches of Sober's Monism." *Journal of Philosophy* 87: 158–161.

Knight, R., Freeland, S., and Landweber, L. (2001). "Rewiring the Keyboard – Evolvability of the Genetic Code." *Nature Reviews: Genetics* 2(1): 49–58.

Krebs, J. and Davies, N. (1981). *An Introduction to Behavioural Ecology*. Sinauer.

Kreitman, M. (1983). "Nucleotide Polymorphism at the Alcohol Dehydrogenase Locus of *Drosophila melanogaster*." *Nature* 304(5925): 412–417.

Kripke, S. (1972/1980). "Naming and Necessity" In D. Davidson, G. Harman and D. Reidel (eds.), *Semantics of Natural Language*. D. Reidel Publishing Co. Expanded version. Harvard University Press.

Lakatos, I. (1978): "Falsification and the Methodology of Scientific Research Programmes." In *The Methodology of Scientific Research Programmes – Philosophical Papers*, vol. 1. Cambridge University Press.

Laland K., Uller, T., Feldman, M. et al. (2015). "The Extended Evolutionary Synthesis – Its Structure, Assumptions and Predictions." *Proceedings of the Royal Society B* 282(1813): 20151019.

Lamarck, J. B. (1809). *Philosophie Zoologique*. 2 volume. Savy. Translated by H. Elliot, *The Zoological Philosophy*. 1914. Macmillan.

Lamotte, M. (1959). "Polymorphism of Natural Populations of *Cepaea Nemoralis*." *Cold Spring Harbor Symposia on Quantitative Biology* 24: 65–84.

Lange, M. and Rosenberg, A. (2011). "Can There Be *A Priori* Causal Models of Natural Selection?" *Australasian Journal of Philosophy* 89(4): 591–599.

Laplace, P. (1812). *A Philosophical Essay on Probability*. Translated by F. Truscott and F. Emory. Dover Publications, 1951.

LaPorte, J. (2004). *Natural Kinds and Conceptual Change*. Cambridge University Press.

Lean, C., Doolittle, F. W., and Bielawska, J. (2022). "Community–Level Evolutionary Processes – Linking Community Genetics with Replicator–Interactor Theory." *Proceedings of the National Academy of Sciences USA* 119(46): e2202538119.

Lederberg, E. and Lederberg, J. (1952). "Replica Plating and Indirect Selection of Bacterial Mutants," *Journal of Bacteriology* 63(3): 399–406.

Lemey, P., Salemi, M., and Vandamme, A.-M (2009). *The Phylogenetic Handbook – A Practical Approach to Phylogenetic Analysis and Hypothesis Testing*, 2nd edition. Cambridge University Press.

Lennox, J. (2001). *Aristotle's Philosophy of Biology*. Cambridge University Press.

Lenski, R. E. and Mittler, J. E. (1993). "The Directed Mutation Controversy and Neo– Darwinism." *Science* 259(5092): 188–194.

Levine, P. (2017). *Eugenics – A Very Short Introduction.* Oxford University Press.

Lewens, T. (2015). *Cultural Evolution – Conceptual Challenges.* Oxford University Press.

Lewin, R. (1980). "Evolutionary Theory under Fire." *Science* 210(4472): 883–887.

Lewis, D. (1973). *Counterfactuals.* Blackwell.

Lewis, D. (1980). "A Subjectivist's Guide to Objective Chance." In R. C. Jeffrey (ed.), *Studies in Inductive Logic and Probability.* Volume II. University of California Press, pp. 263–293.

Lewis, D. (1986). "Postscript to 'Causation.'" In *Philosophical Papers,* volume 2. Oxford University Press, pp. 172–213.

Lewis, D. (1994). "Humean Supervenience Debugged." *Mind* 103: 473–490.

Lewontin, R. (1970a). "Race and Intelligence." *Bulletin of Atomic Scientists* 26(3): 2–8.

Lewontin, R. (1970b). "The Units of Selection." *Annual Review of Ecology and Systematics* 1: 1–18.

Lewontin, R. (1972). "The Apportionment of Human Diversity." *Evolutionary Biology*, 381–398.

Lewontin, R. (1974). "The Analysis of Variance and the Analysis of Causes." *American Journal of Human Genetics* (26): 400–411.

Lewontin, R. and Dunn, L. (1963). "The Evolutionary Dynamics of a Polymorphism in the House Mouse." *Genetics* (45)6: 705–722.

Lewontin, R. and Hubby, J. (1966). "A Molecular Approach to the Study of Genic Heterozygosity in Natural populations. 2. Amount of Variation and Degree of Heterozygosity in Natural Populations of Drosophila *pseudoobscura.*" *Genetics* 54(2): 595–609.

Lewontin, R., Rose, S., and Kamin, L. (1990). *Not in Our Genes – Biology, Ideology and Human Nature.* Penguin Books.

Li, J., Asher, D., Tang, H. et al. (2008). "Worldwide Human Relationships Inferred from Genome – Wide Patterns of Variation." *Science* 319(5866): 1100–1104.

Li, W. H. (1978). "Maintenance of Genetic Variability under the Joint Effect of Mutation, Selection and Random Drift." *Genetics* 90(2): 349–382.

Linnaeus, C. (1758). Systema Naturae. Translated by W. Turton (1806) as *A General System of Nature through the Three Grand Kingdoms of Animals Vegetables and Minerals Systematically Divided into their Several Classes Orders Genera Species and Varieties with their Habitations Manners Economy Structure and Peculiarities.* Lackington, Allen, and Company.

Lloyd, E. (2020). "Units and Levels of Selection." In E. Zalta (ed.), *The Stanford Encyclopedia of Philosophy.* https://plato.stanford.edu/archives/spr2020/entries/selection-units/

Lorenz, E. N. (1972). "Predictability – Does the Flap of a Butterfly's Wings in Brazil Set Off a Tornado in Texas?" *American Association for the Advancement of Science* presentation, December 29.

Lorenz, K. (1963). *Das sogenannte Böse – Zur Naturgeschichte der Aggression.* 1966 English translation *On Aggression.* Harper. 1974 edition.

Luria, S. and Delbrück, M. (1943). "Mutations of Bacteria from Virus Sensitivity to Virus Resistance." *Genetics* 28(6): 491–511.

Lyell, C. (1830–1833). *The Principles of Geology, Being an Attempt to Explain the Former Changes of the Earth's Surface, by Reference to Causes Now in Operation.* Volumes 1,2,3. Murray.

Malet, K. (2007). "Species, Concepts of." *Encyclopedia of Biodiversity.* DOI: 10.1016/B978–0–12–822562–2.00022–0.

Malthus, T. (1798). *An Essay on the Principle of Population.* A.M. Kelly, 1965.

Matthen, M. and Ariew, A. (2002). "Two Ways of Thinking about Fitness and Natural Selection." *Journal of Philosophy* 119: 55–83.

Mannouris, C. (2011). "Darwin's 'Beloved Barnacles' – Tough Lessons in Variation." *History and Philosophy of the Life Sciences* 33(1): 51–70.

Marouli, E., Graff, M., Medina-Gomez, C. et al. (2017). "Rare and Low– Frequency Coding Variants Alter Human Adult Height." *Nature* 542(7640): 186–190.

Maudlin, T. (2019). *Philosophy of Physics – Quantum Mechanics.* Princeton University Press.

Maxwell, M. (2024). "Can Math Save the Propensity Interpretation of Fitness?" unpublished.

Maxwell, M. and Sober, E. (2023a). "Gradualism, Natural Selection, and the Randomness of Mutation – Fisher, Kimura, and Orr, Connecting the Dots." *Biology and Philosophy.*

Maxwell, M. and Sober, E. (2023b). "When Does Drift Dominate Selection? A Zeroing-Out Criterion." unpublished.

Maynard Smith, J. (1982). *Evolution and the Theory of Games.* Cambridge University Press.

Maynard Smith, J. (1989). "The Causes of Extinction." *Philosophical Transactions of the Royal Society London B* 325(1228): 241–252.

Maynard Smith, J. (1991). "Byerly and Michod on Fitness." *Biology and Philosophy* 6(1): 37.

Maynard Smith, J. and Haigh, J. (1974). "The Hitch–hiking Effect of a Favourable Gene." *Genetics Research* 23(1): 23–35.

Maynard Smith, J. and Price, G. (1973). "The Logic of Animal Conflict." *Nature* 246(5427): 15–18.

Mayr, E. (1942). *Systematics and the Origin of Species.* Columbia University Press.

Mayr, E. (1959). "Typological versus Population Thinking." In J. Meggers (ed.), *Evolution and Anthropology: A Centennial Appraisal*. The Anthropological Society of Washington, pp. 1–10. Reprinted in Mayr (1976).

Mayr, E. (1963). *Animal Species and Evolution*. Harvard University Press.

Mayr, E. (1970). *Populations, Species, and Evolution*. Harvard University Press.

Mayr, E. (1976). *Evolution and the Diversity of Life*. Harvard University Press.

Mayr, E. (1982). *The Growth of Biological Thought: Diversity, Evolution, and Inheritance*. Harvard University Press.

Mayr, E. (2007). "Darwin's Five Theories of Evolution" In *What Makes Biology Unique?* Cambridge University Press, pp. 97–115.

McShea, D. and Brandon, R. (2010). *Biology's First Law*. University of Chicago Press.

Meunier, R. (2022). "Gene." In E. Zalta (ed.), *The Stanford Encyclopedia of Philosophy*. https://plato.stanford.edu/archives/fall2022/entries/gene/

Mikkola, M. (2022). "Feminist Perspectives on Sex and Gender." In E. Zalta and U. Nodelman (eds.), *The Stanford Encyclopedia of Philosophy*. https://plato.stanford.edu/archives/fall2023/entries/feminism-gender/

Miller, D. W. (1994). *Critical Rationalism – A Restatement and Defence*. Open Court.

Mills, S. and Beatty, J. (1979). "The Propensity Interpretation of Fitness." *Philosophy of Science* 46(2): 263–288.

Millstein, R. (2002a). "Are Random Drift and Natural Selection Conceptually Distinct?" *Biology and Philosophy* 17(1): 33–53.

Millstein, R. (2002b). "How not to Argue for the Indeterminism of Evolution – A Look at Two Recent Attempts to Settle the Issue." Unpublished manuscript.

Millstein, R. (2006). "Natural Selection as a Population-Level Causal Process." *British Journal for the Philosophy of Science* 57(4): 627–653.

Millstein, R. (2009). "Populations as Individuals." *Biological Theory* 4(3): 267–273.

Millstein, R. (2021). "Genetic Drift." In E. Zalta (ed.), *The Stanford Encyclopedia of Philosophy*. https://plato.stanford.edu/archives/spr2021/entries/genetic-drift/

Mishler, B. and Wilkins, J. (2008). "The Hunting of the SNaRC – A Snarky Solution to the Species." *Philosophy, Theory, and Practice in Biology* 10(1).

Mitchell, W. and Valone, T. (1990). "The Optimization Research Program – Studying Adaptations by their Function." *Quarterly Review of Biology* 65(1): 43–52.

Monroe, J., Srikant, T., Carbonell–Bejerano, P. et al. (2022). "Mutation Bias Reflects Natural Selection in *Arabidopsis thaliana*." *Nature* 602: 101–105.

Moutinho, A., Eyre-Walker, A., and Dutheil, J. (2022). "Strong Evidence for the Adaptive Walk Model of Gene Evolution in Drosophila and Arabidopsis. *PLoS Biology* 20(9): e3001775. https://doi.org/10.1371/journal.pbio.3001775.

Muirhead, C. and Presgraves, D. (2021). “Satellite DNA-mediated Diversification of a *Sex-Ratio* Meiotic Drive Gene Family in *Drosophila.*” *Nature Ecology and Evolution* 5(12): 1604–1612.

Müller–Wille, S. (2007). “Collection and Collation – Theory and Practice of Linnaean Botany.” *Studies in History and Philosophy of Biological and Biomedical Science* 38(3): 541–562.

Novembre, J. (2022). “The Background and Legacy of Lewontin’s Apportionment of Human Genetic Diversity.” *Philosophical Transactions of the Royal Society B* 368: 20120404. doi: 10.1098/rstb.2012.0404.

Ohta, T. (1973). “Slightly Deleterious Mutant Substitutions in Evolution.” *Nature* 246(5428): 96–98.

Okasha, S. (2002). “Darwinian Metaphysics – Species and the Question of Essentialism.” *Synthese* (131): 191–213.

Okasha, S. (2006). *Evolution and the Levels of Selection*. Oxford University Press.

Okasha, S. (2020). “Biological Altruism.” In E. Zalta (ed.), *The Stanford Encyclopedia of Philosophy*. https://plato.stanford.edu/archives/sum2020/entries/altruism-biological/

Okasha S. and Otsuka J. (2020). “The Price Equation and the Causal Analysis of Evolutionary Change.” *Philosophical Transactions of the Royal Society B* 375: 20190365. http://dx.doi.org/10.1098/rstb.2019.0365

Orr, H. A. (1998). “The Population Genetics of Adaptation – The Distribution of Factors Fixed during Adaptive Evolution.” *Evolution* 52(4): 935–949.

Orr, H. A. (2000). “Adaptation and the Cost of Complexity.” *Evolution* 54(1): 13–20.

Orr, H. A. (2009). “Darwin and Darwinism – The (Alleged) Social Implications of *The Origin of Species*.” *Genetics* 183(3): 767–772.

Orr, H. A. and Coyne, J. A. (1992). “The Genetics of Adaptation – A Reassessment.” *The American Naturalist* 140(5): 725–742.

Orzack, S. and Sober, E. (1994). “Optimality Models and the Test of Adaptationism.” *American Naturalist* 143(3): 361–380.

Orzack, S. and Sober, E. (2001). “Adaptation, Phylogenetic Inertia, and the Method of Controlled Comparisons.” In S. Orzack and E. Sober (eds.), *Adaptationism and Optimality*. Cambridge University Press, pp. 45–63.

Ospovat, D. (1981). *The Development of Darwin’s Theory*. Cambridge University Press.

Owen, R. (1849). *On the Nature of Limbs*. University of Chicago Press, 2007.

Padian, K. (1999). “Charles Darwin’s Views of Classification in Theory and Practice.” *Systematic Biology* 48(2): 352–364.

Paley, W. (1802). *Natural Theology or Evidences of the Existence and Attributes of the Deity*. Faulder.

Park, J. H., Wacholder, S., Gail, M. H. et al. (2010). "Estimation of Effect Size Distribution from Genome – Wide Association Studies and Implications for Future Discoveries." *Nature Genetics*, 42(7): 570–575.

Parker, G. (1978). "Searching for Mates." In J. Krebs and N. Davies (eds.), *Behavioral Ecology – An Evolutionary Approach.* Blackwells, pp. 214–244.

Parker, G., Simmons, L., Stockley, P., McChristie, D., and Charnov, E. (1999). "Optimal Copula Duration in Yellow Dung Flies – Effects of Female Size and Egg Content." *Animal Behaviour* 57(4): 795–805.

Paul, D. (1995). *Controlling Human Heredity – 1865 to the Present.* Humanities Press.

Pawitan, Y. (2001). *In All Likelihood – Statistical Modelling and Inference Using Likelihood.* Oxford University Press.

Pearl, J. (2000). *Causality – Models, Reasoning, and Inference*. Cambridge University Press.

Pedroso, M. (2014). "Origin Essentialism in Biology." *The Philosophical Quarterly* 64(254): 60–81.

Pence, C. and Ramsey, G. (2013). "A New Foundation for the Propensity Definition of Fitness." *British Journal for the Philosophy of Science* 64: 851–881.

Pence, C. and Ramsey, G. (2015). "Is Organismic Fitness at the Basis of Evolutionary Theory?" *Philosophy of Science* 82(5): 1081–1091.

Peterson, E. (forthcoming). *The Shortest History of Eugenics.* New York: The Experiment.

Plomin, R., DeFries, J. C., McClearn, G. E., and McGuffin, P. (2017). *Behavioral Genetics: A Primer*, 2nd edition. W.H. Freeman.

Polger, T. and Shapiro, L. (2016). *The Multiple Realization Book.* Oxford University Press.

Popper, K. (1959). *The Logic of Scientific Discovery.* Routledge.

Popper, K. (1990). *A World of Propensities.* Thoemmes.

Posada, D. and Buckley, T. R. (2004). "Model Selection and Model Averaging in Phylogenetics – Advantages of Akaike Information Criterion and Bayesian Approaches over Likelihood Ratio Tests." *Systematic Biology* 53(5): 793–808.

Posada D. and Crandall K. (2001). "Evaluation of Methods for Detecting Recombination from DNA Sequences: Computer Simulations." *Proceedings of the National Academies of Science USA* 98(24): 13757–13762.

Poundstone, W. (1993). *Prisoner's Dilemma – John von Neumann, Game Theory, and the Puzzle of the Bomb.* Anchor.

Price, G. R. (1970). "Selection and Covariance." *Nature.* 227(5257): 520–521.

Price, G. R. (1972). "Extension of Covariance Selection Mathematics." *Annals of Human Genetics* 35(4): 485–490.

Quine, W. (1953). "Two Dogmas of Empiricism." In *From a Logical Point of View.* Harvard University Press, pp. 20–46.

Ramsey, G. (2006). "Block Fitness." *Studies in History and Philosophy of Biological and Biomedical Sciences* 37: 484–498.

Rawls, J. (1970). *A Theory of Justice*. Harvard University Press.

Reichenbach, H. (1956). *The Direction of Time*. University of California Press.

Reichenbach, H. (1958). *The Philosophy of Space and Time*. University of California Press.

Reilly, P. R. (1991). *The Surgical Solution – A History of Involuntary Sterilization in the United States*. Johns Hopkins University Press.

Reisman, K. and Forber, P. (2005). "Manipulation and the Causes of Evolution." *Philosophy of Science* 72(5): 1115–1125.

Rice, S. (2004). *Evolutionary Theory – Mathematical and Conceptual Foundations*. Sinauer.

Richard, R. (1997). "Darwin and the Inefficacy of Artificial Selection." *Studies in History and Philosophy of Science* 28(1): 75–97.

Ridley, M. (1986). *Evolution and Classification – The Reformation of Cladism*. Longman.

Rieseberg, L. H. and Brouillet, L. (1994). "Are Many Plant Species Paraphyletic?" *Taxon* 43(1): 21–32.

Roche, W. and Sober, E. (2019). "Observation Selection Effects and Discrimination Conduciveness." *Philosophical Imprint* 19(40): 1–26.

Rosenberg, N., Pritchard, J., Weber, J. et al. (2002). "Genetic Structure of Human Populations." *Science* 298(5602): 2381–2384.

Roughgarden, J. (1979). *Theory of Population Genetics and Evolutionary Ecology*. Prentice Hall.

Rousseau, J. J. (1755). *Discourse on Inequality*. Penguin, 1985.

Royall, R. (1997). *Statistical Evidence – A Likelihood Paradigm*. Chapman and Hall.

Ruse, M. (1979). *The Darwinian Revolution – Science Red in Tooth and Claw*. University of Chicago Press.

Salmon, W. (1984). *Scientific Explanation and the Causal Structure of the World*. Princeton University Press.

Schaffer, J. (2007). "Deterministic chance?" *The British Journal for the Philosophy of Science* 58(2): 113–140.

Scerri, E. (2007). *The Periodic Table – Its Story and Its Significance*. Oxford University Press.

Shanahan, T. (2011). "Phylogenetic Inertia and Darwin's Higher Law." *Studies in History and Philosophy of Biological and Biomedical Sciences* 42(1): 60–68.

Shapiro, L. and Sober, E. (2007). "Epiphenomenalism – The Do's and the Don'ts." In P. Machamer and G. Wolters (eds.), *Thinking about Causes*. University of Pittsburgh Press, pp. 235–264.

Sidelle, A. (1989). *Necessity, Essence and Individuation – A Defence of Conventionalism*. Cornell University Press.

Sidelle, A. (2009). "Conventionalism and the Contingency of Conventions." *Nous* 43(2): 224–241.

Simons, Y. B., Bullaughey, K., Hudson, R. R., and Sella, G. (2018). "A Population Genetic Interpretation of GWAS Findings for Human Quantitative Traits." *PLoS Biology* 16(3). https://doi.org/ 10.1371/journal.pbio.2002985.

Simpson, E. H. (1951). "The Interpretation of Interaction in Contingency Tables." *Journal of the Royal Statistical Society B* 13(2): 238–241.

Simpson, G. G. (1945). *The Principles of Classification and a Classification of Mammals*. Vol. 85. American Museum of Natural History.

Simpson, G. G. (1967). *The Meaning of Evolution – A Study of the History of Life and of Its Significance for Man*, Revised edition. Yale University Press.

Skyrms, B. (2004). *The Stag Hunt and the Evolution of Social Structure*. Cambridge University Press.

Sly, A. (2011). "Reconstruction for the Potts Model." *Annals of Probability* 39(4): 1365–1406.

Smith, E. and Winterhalder, B. (1992). *Evolutionary Ecology and Human Behavior*. Aldine de Gruyter.

Sneath, P. and Sokal, R. (1973). *Numerical Taxonomy – The Principles and Practice of Numerical Classification*. W.H. Freeman.

Sober, E. (1980a). "Evolution, Population Thinking, and Essentialism." *Philosophy of Science* 47(3): 350–383.

Sober, E. (1980b). "Holism, Individualism, and the Units of Selection." *PSA – Proceedings of the Biennial Meeting of the Philosophy of Science Association* 2: 93–121.

Sober, E. (1984). *The Nature of Selection*. MIT Press.

Sober, E.(1988a). "Apportioning Causal Responsibility." *Journal of Philosophy* 85(6): 303–318.

Sober, E. (1988b). "Likelihood and Convergence." *Philosophy of Science* 55: 228–237.

Sober, E. (1988c). *Reconstructing the Past – Parsimony, Evolution, and Inference*. MIT Press.

Sober, E. (1989). "Independent Evidence about a Common Cause." *Philosophy of Science* 56(2): 275–287.

Sober, E. (1990). *Core Questions in Philosophy*. Macmillan. 8th edition 2021, Taylor & Francis.

Sober, E. (1991a). "Models of Cultural Evolution." In P. Griffiths (ed.), *Trees of Life – Essays in Philosophy of Biology*. Kluwer, pp. 17–38.

Sober, E. (1991b). "Organisms, Individuals, and Units of Selection." In A. Tauber (ed.), *Organism and the Origin of Self*. Kluwer, pp. 273–296.

Sober. E. (1992). "The Evolution of Altruism – Correlation, Cost, and Benefit." *Biology and Philosophy* 7: 177–188.

Sober, E. (1993). *Philosophy of Biology*. Westview Press.

Sober, E. (1997). "Two Outbreaks of Lawlessness in Recent Philosophy of Biology." *Philosophy of Science* 64(4): S458–S467.

Sober, E. (1998). "Three Differences between Evolution and Deliberation." In P. Danielson (ed.), *Modeling Rationality, Morality, and Evolution*. Oxford University Press, pp. 408–422.

Sober, E. (1999). "The Multiple Realizability Argument against Reductionism." *Philosophy of Science* 66(4): 542–564.

Sober, E. (2001a). "The Two Faces of Fitness." In R. Singh, D. Paul, C. Krimbas, and J. Beatty (eds.), *Thinking about Evolution – Historical, Philosophical, and Political Perspectives*. Cambridge University Press, vol. 2, pp. 309–321.

Sober, E. (2001b). "Venetian Sea Levels, British Bread Prices, and the Principle of the Common Cause." *British Journal for the Philosophy of Science* 52(2): 331–346.

Sober, E. (2003). "Contingency or Inevitability – What Would Happen if the Evolutionary Tape Were Replayed?" *New York Times*, November 30.

Sober, E. (2004). "The Contest between Likelihood and Parsimony." *Systematic Biology* 53(4): 6–16.

Sober, E. (2008). *Evidence and Evolution*. Cambridge University Press.

Sober, E. (2010). "Evolutionary Theory and the Reality of Macro-Probabilities." In E. Eells and J. Fetzer (eds.), *The Place of Probability in Science*. Springer, pp. 133–162.

Sober, E. (2011a). "*A Priori* Causal Models of Natural Selection." *Australasian Journal of Philosophy* 89(4): 571–589.

Sober, E. (2011b). *Did Darwin Write the Origin Backwards? Philosophical Essays on Darwin's Theory*. Prometheus Books.

Sober, E. (2011c). "Realism, Conventionalism, and Causal Decomposition in Units of Selection – Reflections on Samir Okasha's Evolution and the Levels of Selection." *Philosophy and Phenomenological Research* 82(1): 221–231.

Sober, E. (2013). "Trait Fitness Is Not a Propensity, but Fitness Variation Is." *Studies in History and Philosophy of Biological and Biomedical Sciences* 44: 336–341.

Sober, E. (2014). "Evolutionary Theory, Causal Completeness, and Theism – The Case of 'Guided' Mutation." In D. Walsh and P. Thompson (eds.), *Evolutionary Biology – Conceptual, Ethical, and Religious Issues*. Cambridge University Press, pp. 31–44.

Sober, E. (2015). *Ockham's Razors – A User's Manual*. Cambridge University Press.

Sober, E. (2018). *The Design Argument*. Cambridge University Press.

Sober, E. and Steel, M. (2002). "Testing the Hypothesis of Common Ancestry." *Journal of Theoretical Biology* 218(4): 395–408.

Sober, E. and Steel, M. (2011). "Entropy Increase and Information Loss in Markov Models of Evolution." *Biology and Philosophy* 26(2): 223–250.

Sober, E. and Steel, M. (2014). “Time and Knowability in Evolution.” *Philosophy of Science* 81(4): 558–579.

Sober, E. and Steel, M. (2015). “How Probable Is Common Ancestry According to Different Evolutionary Processes?” *Journal of Theoretical Biology* 373: 111–116.

Sober, E. and Steel. M. (2017). “Similarities as Evidence for Common Ancestry – A Likelihood Epistemology.” *British Journal for the Philosophy of Science* 68(3): 617–638.

Sober, E. and Wilson, D. S. (1998). *Unto Others – The Psychology and Evolution of Unselfish Behavior.* Harvard University Press.

Sober, E. and Wilson, D. S. (2011). “*Adaptation and Natural Selection* Revisited.” *Journal of Evolutionary Biology* 24(2): 462–468.

Sokal, R. and Sneath, P. (1963). *Principles of Numerical Taxonomy.* W.H. Freeman.

Spencer, H. (1864). *Principles of Biology.* Williams and Norgate.

Spencer, Q. (2014). “A Radical Solution to the Race Problem.” *Philosophy of Science* 81(5): 1025–1038.

Stanley, S. (1975). “A Theory of Evolution Above the Species Level.” *Proceedings of the National Academy of Sciences USA* 72(2): 646–650.

Stanley, S. (1979). *Macroevolution – Pattern and Process.* W. H. Freeman.

Steel, M. and Penny, D. (2000). “Parsimony, Likelihood, and the Role of Models in Molecular Phylogenetics.” *Molecular Biology and Evolution* 17: 839–850.

Stephens, C. (2004). “Selection, Drift, and the ‘Forces’ of Evolution.” *Philosophy of Science* 71(4): 550–570.

Sterelny, K. and Griffiths, P. (1999). *Sex and Death.* University of Chicago Press.

Sterelny, K. and Kitcher, P. (1988). “The Return of the Gene.” *Journal of Philosophy* 85: 339–360.

Sturtevant, A. H. (1937). “On the Effects of Selection on Mutation Rate.” *Quarterly Review of Biology* 12(4): 464–467.

Swenson, W., Wilson, D. S., and Elias, R. (2000). “Artificial Ecosystem Selection.” *Proceedings of the National Academy of Sciences* 97: 9110–9114.

Tang, H., Peng, J., Wang, P., and Risch, N. (2005). “Estimation of Individual Admixture: Analytical and Study Design Considerations.” *Genetic Epidemiology* 28(4): 289–301.

Trivers, R. (1971). “The Evolution of Reciprocal Altruism.” *The Quarterly Review of Biology* (46)1: 35–57.

True, J. and Haag, E. (2001). “Developmental System Drift and Flexibility in Evolutionary Trajectories.” *Evolutionary Development* 3(2): 109–119.

Tuffley, C. and Steel, M. (1997). “Links between Maximum Likelihood and Maximum Parsimony Under a Simple Model of Site Substitution.” *Bulletin of Mathematical Biology* 59(3): 581–607.

Uhr G., Dohnalová, L., and Thaiss, C. (2019). “The Dimension of Time in Host-Microbiome Interactions.” *mSystems* 4(1): e00216–e00218.

Van Fraassen, B. (1980). *The Scientific Image*. Oxford University Press.

Van Fraassen, B. (1989). *Laws and Symmetry*. Oxford University Press.

Van Valen, L. (1973). "A New Evolutionary Law." *Evolutionary Theory* 1: 1–30.

Vassend, O., Sober, E., and Fitelson, B. (2017). "The Philosophical Significance of Stein's Paradox." *European Journal for the Philosophy of Science* 7(3): 411–433.

Velasco, J. (2008). "Species Concepts Should Not Conflict with Evolutionary History, But Often Do." *Studies in the History and Philosophy of Biology and Biomedical Sciences*. 39(4): 407–441.

Velasco, J. (2018). "Universal Common Ancestry, LUCA, and the Tree of Life – Three Distinct Hypotheses about the Evolution of Life." *Biology and Philosophy* 33(5): 1–18.

Veuille, M. (2019). "Chance, Variation and Shared Ancestry – Population Genetics After the Synthesis." *Journal of the History of Biology* 52: 537–567.

Vigen, T. (2015). *Spurious Correlations*. Hachette.

Visscher, P. M., McEvoy, B., and Yang, J. (2010). "From Galton to GWAS: Quantitative Genetics of Human Height." *Genetics Research* 92(5–6): 371–379.

Von Neumann, J. and Morgenstern, O. (1944). *Theory of Games and Economic Behavior*. Princeton University Press.

Vorzimmer, P. (1969). "Darwin, Malthus, and the Theory of Natural Selection." *Journal of the History of Ideas* 30(4): 527–542.

Vrba, E. (1980). "Evolution, Species, and Fossils – How Does Life Evolve?" *South African Journal of Science* 76(2): 61–84.

Walsh, D. (2000). "Chasing Shadows – Natural Selection and Adaptation." *Studies in the History and Philosophy of Biology and the Biomedical Sciences* 31: 135–153.

Walsh, D. (2004). "Bookkeeping or Metaphysics? The Units of Selection Debate." *Synthese* 138: 337–361.

Walsh, D., Lewens, T., and Ariew, A. (2002). "The Trials of Life – Natural Selection and Random Drift." *Philosophy of Science* 69: 452–473.

Waters, K. (1991). "Tempered Realism about Units of Selection." *Philosophy of Science* 58: 553–573.

Waters, K. (2003). "The Arguments in Darwin's *Origin*." In J. Hodge and G. Radick (eds.), *Cambridge Companion to Darwin*. Cambridge University Press, pp. 116–139.

Waters, K. (2005). "Why Genic and Multilevel Selection Theories Are Here to Stay." *Philosophy of Science* 72: 311–333.

Watson, H. and Galton, F. (1875). "On the Probability of the Extinction of Families." *Journal of the Anthropological Institute of Great Britain and Ireland* 4: 138–144.

Weismann, A. (1892). *Das Keimplasma – eine Theorie der Vererbung*. Fischer. Germ–Plasm, a Theory of Heredity (English translation), Charles Scribner's Sons, 1893.

Whewell, W. (1840). *The Philosophy of the Inductive Sciences – Founded Upon Their History*. Excerpted in *William Whewell's Theory of Scientific Method*. R. Butts (ed.). University of Pittsburgh Press, 1968.

Wiley, E. (1981). *Phylogenetics – The Theory and Practice of Phylogenetic Systematics*. Wiley.

Williams, G. C. (1957). "Pleiotropy, Natural Selection, and the Evolution of Senescence." *Evolution* 11(4): 398–411.

Williams, G. C. (1966). *Adaptation and Natural Selection*. Princeton University Press.

Williams, G. C. and Williams, D. C. (1957). "Natural Selection of Individually Harmful Social Adaptations Among Sibs with Special Reference to Social Insects." *Evolution* 11(1): 32–39.

Wilson, D. S. (1975). "A Theory of Group Selection." *Proceedings of the National Academy of Sciences USA* 72(1): 143–146.

Wimsatt, W. (1980). "Reductionistic Research Strategies and Their Biases in the Units of Selection Controversy." In T. Nickles (ed.), *Scientific Discovery – Case Studies*. D. Reidel Publishing Co., pp. 213–259.

Winsor, M. (2006a). "Linnaeus's Biology Was Not Essentialist." *Annals of the Missouri Botanical Garden* 93(1): 2–7.

Winsor, M. (2006b). "The Creation of the Essentialism Story – An Exercise in Metahistory." *History and Philosophy of the Life Sciences* 28(2): 149–174.

Winther, R. (2023). *Our Genes – A Philosophical Perspective on Human Evolutionary Genomics*. Cambridge University Press.

Wittgenstein, L. (1953). *Philosophical Investigations*. Basil Blackwell.

Woese, C. (1998). "The Universal Ancestor." *Proceedings of the National Academy of Sciences USA* 95(12): 6854–6859.

Woese, C. (2000). "Interpreting the Universal Phylogenetic Tree." *Proceedings of the National Academy of Sciences USA* 97(15): 8392–8396.

Wolfe, K., Sharp, P., and Li, W. (1989). "Mutation Rates Differ Among Regions of the Mammalian Genome." *Nature* 337(6204): 283–285.

Woodward, J. (2005). *Making Thing Happen*. Oxford University Press.

Wright, E., Levine, A., and Sober, E. (1993). *Reconstructing Marxism – Essays on Explanation and the Theory of History*. Verso Books.

Wright, L. (1973). "Functions." *Philosophical Review* 82(2): 139–168.

Wright, S. (1922). "Coefficients of Inbreeding and Relationship." *American Naturalist* 56(645): 330–338.

Wright, S. (1931). "Evolution in Mendelian Populations." *Genetics* 16(2): 97–159.

Wright, S. (1937). "The Distribution of Gene Frequencies in Populations." *Proceedings of the National Academy of Sciences* 23(6): 307–320.

Yule, G. U. (1926). "Why Do We Sometimes Get Nonsense – Correlations between Time Series? A Study in Sampling and the Nature of Time Series." *Journal of the Royal Statistical Society* 89(1): 1–63.

Zuckerkandl, E. and Pauling, L. (1965). "Evolutionary Divergence and Convergence in Proteins." In V. Brysson and H. Vogel (eds.), *Evolving Genes and Proteins.* Academic Press, pp. 97–166.

Index

Printed in the United States
by Baker & Taylor Publisher Services

Printed in the United States
by Baker & Taylor Publisher Services